About IFPRI

The International Food Policy Research Institute (IFPRI), a research center of CGIAR, provides research-based policy solutions to sustainably reduce poverty and end hunger and malnutrition in low- and middle-income countries. IFPRI was established in 1975 to identify and analyze alternative national and international strategies and policies for meeting the food needs of the developing world, with particular emphasis on low-income countries and on the poorer groups in those countries. Partnerships, communications, capacity strengthening, and data and knowledge management are essential components for translating IFPRI's research to action and impact. The Institute's regional and country programs play a critical role in responding to demand for food policy research and in delivering holistic support to country-led development. IFPRI collaborates with partners around the world.

www.ifpri.org

About CGIAR

CGIAR is a global partnership for a food-secure future dedicated to transforming food, land, and water systems in a climate crisis. CGIAR science is dedicated to reducing poverty, enhancing food and nutrition security, and improving natural resources and ecosystem services. As the world's largest agricultural innovation network, its research is carried out by 15 CGIAR Centers working around the world in close collaboration with hundreds of partners, including national and regional research institutes, civil society organizations, academia, development organizations, and the private sector.

www.cgiar.org

IFPRI

GLOBAL FOOD POLICY REPORT 2023

RETHINKING FOOD CRISIS RESPONSES

gfpr.ifpri.info

A Peer-Reviewed Publication

RECOMMENDED CITATION: International Food Policy Research Institute. 2023. *2023 Global Food Policy Report: Rethinking Food Crisis Responses*. Washington, DC: International Food Policy Research Institute. https://doi.org/10.2499/9780896294417

International Food Policy Research Institute
1201 Eye Street, NW
Washington, DC 20005-3915 USA
www.ifpri.org

ISBN: 978-0-89629-441-7
ISSN: 2329-2873
DOI: https://doi.org/10.2499/9780896294417
Cataloging-in-Publication Data is available from the Library of Congress.

Photo credits
COVER: Lumiereist / Shutterstock.com.
CHAPTER IMAGES: p. 6 hikrcn / Shutterstock.com; p. 20 Bumble Dee / Shutterstock.com; p. 34 Martin Mecnarowski / Shutterstock.com; p. 36 Abdul Majeed / European Union; p. 44 Andre Nery / Shutterstock.com; p. 52 Dominic Chavez / World Bank; p. 62 Riccardo Mayer / Shutterstock.com; p. 72 ymphotos / Shutterstock.com; p. 82 fivepointsix/ Shutterstock.com.

BOOK LAYOUT: Jason Chow
EDITORIAL MANAGER: Pamela Stedman-Edwards

Contents

Foreword

The past decade has been marked by multiple, often overlapping, crises. The COVID-19 pandemic, various natural disasters, and the ongoing war in Ukraine have all threatened the fabric of our global food systems. This string of crises has left an indelible mark. In too many places, progress in reducing poverty and malnutrition has been reversed, with long-term implications for people's health and livelihoods. While in some ways our food systems have also proved surprisingly resilient, as both the private and public sectors have stepped up to meet new needs, the challenges are huge. As climate change worsens and geopolitical strife grows amid the threat of more frequent pandemics, crises may well become more common and more devastating.

Now is the time to rethink how we address food crises. Many governments, donors, and international organizations have called for moving beyond humanitarian responses that are implemented only after a crisis begins, toward better prediction, preparation, and resilience building that will make future crises less devastating. While discussions of a humanitarian-development-peace approach have been underway for years, this report aims to provide a solid policy basis for moving forward.

The *2023 Global Food Policy Report* explores a growing body of evidence on how diverse policy responses can reduce both the immediate and longer-term impacts of food crises, and improve livelihoods, incomes, and food security and nutrition for the future. Drawing on research from IFPRI and other CGIAR centers, it provides evidence-based policy recommendations for governments, donors, and nongovernmental organizations.

We hope that this year's report will help shape a transformation in how we respond to the shocks that threaten our food systems, by contributing evidence and policy options to inform discussions among local, national, and global policymakers. We look forward to engaging with many partners around the world to expand this research work and support action for better crisis response.

JOHAN SWINNEN
Director General, IFPRI
Managing Director, Systems Transformation, CGIAR

Acknowledgments

The *2023 Global Food Policy Report* was prepared under the overall leadership of Johan Swinnen, Katrina Kosec, and a core team comprising Charlotte Hebebrand, Pamela Stedman-Edwards, Sivan Yosef, Claire Davis, and Jamed Falik.

Contributions were made by Kibrom Abay, Kamiljon Akramov, Ben Belton, Samuel Benin, Lynn Brown, Kevin Chen, Eugenio Díaz-Bonilla, Xinshen Diao, Olivier Ecker, Friederike Greb, Manuel Hernandez, Kalle Hirvonen, Arif Husain, Anjani Kumar, Sikandra Kurdi, David Laborde, Peter Läderach, Jessica Leight, Hazel Malapit, Rui Mao, Wim Marivoet, Harriet Mawia, Jean-François Maystadt, Brian McNamara, Bart Minten, Valeria Piñeiro, Miriam Raouf, Shahidur Rashid, Thomas Reardon, Brendan Rice, Sandra Ruckstuhl, John Ulimwengu, Rob Vos, Sivan Yosef, and Yunyi Zhou.

Production of the report was led by Pamela Stedman-Edwards. Jason Chow was responsible for design and layout. Editorial assistance was provided by Claire Davis and Gillian Hollerich.

We would like to thank IFPRI's donors, including all those who supported IFPRI's research through their contributions for the CGIAR Fund: https://cgiar.org/funders/

CHAPTER 1

The Road to Resilience
Rethinking Responses to Food Crises

JOHAN SWINNEN AND KATRINA KOSEC

Johan Swinnen is managing director, Systems Transformation, CGIAR, and director general, International Food Policy Research Institute (IFPRI). **Katrina Kosec** is a senior research fellow, Development Strategies and Governance Unit, IFPRI.

KEY MESSAGES

Calls to rethink responses to food crises have arisen from recent overlapping shocks to food systems – including the COVID-19 pandemic, increased food prices, conflicts, and natural disasters – and from concerns that crises are becoming more frequent, complex, and protracted. Now is an opportune moment to develop more permanent responses to food crises, guided by strong evidence on the impact of policies, programming, tools, and governance approaches. Drawing on research from IFPRI and colleagues, this report provides a broad set of evidence-based recommendations for better predicting and preparing for crises, addressing crises when they occur, and building equity and the resilience of food systems.

- Early warning systems can facilitate preemptive, rapid, and context-appropriate responses, provided they are well coordinated and based on frequent monitoring of key indicators and understanding of how structural risks can aggravate shocks to food security.
- Anticipatory action frameworks, which help prepare and organize humanitarian aid before crises strike, show promise both for mitigating crises and supporting long-term development efforts.
- Agrifood value chains can support livelihoods and food security during crises when governments maintain a business environment that fosters flexibility and technical and financial innovation, and provide essential infrastructure and targeted assistance for at-risk value chain actors.
- Social protection systems are essential to reducing the impact of crises; they can build resilience prior to a crisis and facilitate recovery when they are flexible, shock-responsive, and carefully targeted. Integrating social protection with gender and climate goals can further empower women and promote sustainability.
- Improvements in collecting gender-disaggregated data, particularly amid crises, and tracking progress toward clear gender targets can promote gender equality. Likewise, including women's voices in policymaking and programming decisions can help ensure that crisis responses improve rather than erode gender equality.
- Forced migration can create both challenges and opportunities for development. Migrants can provide benefits for both the host and sending communities when policies facilitate their integration into host communities and support those who remain.
- The resilience of food systems depends critically on good governance; governance determines the ability to implement and sustain effective policies and programming to offset negative shocks, curb incentives for violent conflict, and support the functioning of markets and private sector investments.
- Recent events have highlighted the need for crisis response funding to be expanded and used more efficiently. Repurposing agricultural support funds and better leveraging private sector funds could bolster investment in long-term resilience.

In 2022, the world faced multiple crises. Globally, disruptions to food systems continued amid a protracted pandemic, major natural disasters, civil unrest and political instability, and the growing impacts of climate change, all while the war in Ukraine exacerbated a global food and fertilizer crisis. Yet some aspects of food systems have proved surprisingly resilient in the face of crisis. The International Food Policy Research Institute's (IFPRI's) *2021 Global Food Policy Report: Transforming Food Systems after COVID-19* showed, for example, that adopting new business models helped to keep food value chains functioning during the pandemic, and expanding social protection programs reduced negative impacts on food security.

Moving forward, a range of promising approaches have already been identified to promote resilience along with other development goals. For example, IFPRI's *2022 Global Food Policy Report: Climate Change & Food Systems* outlines several policies, such as rural access to clean energy, trade reforms, and landscape governance, that address climate change while also supporting poverty reduction and food security. While these advances hold potential, the global community still needs a better understanding of how food systems and their various actors respond to crises, and which policy interventions could successfully support households and food value chains in different countries and crisis contexts.

Over many years, IFPRI has built a wealth of evidence on policies, programming, tools, and approaches that reduce hunger and poverty and promote sustainable development and women's empowerment, including during crises. With this report, we present some of our most recent research in response to the growing call for a more holistic approach to preparing for, detecting, averting, mitigating, and responding to crises. Heeding this call will require a shift from simply responding to crises with humanitarian assistance to a concerted approach that strengthens the humanitarian-development-peace nexus, supports and empowers the most vulnerable, and builds more resilient food systems for the future.

In this first chapter, we highlight the vulnerability of food systems to frequent and damaging shocks

FIGURE 1 Prevalence and number of undernourished worldwide, 2000–2021

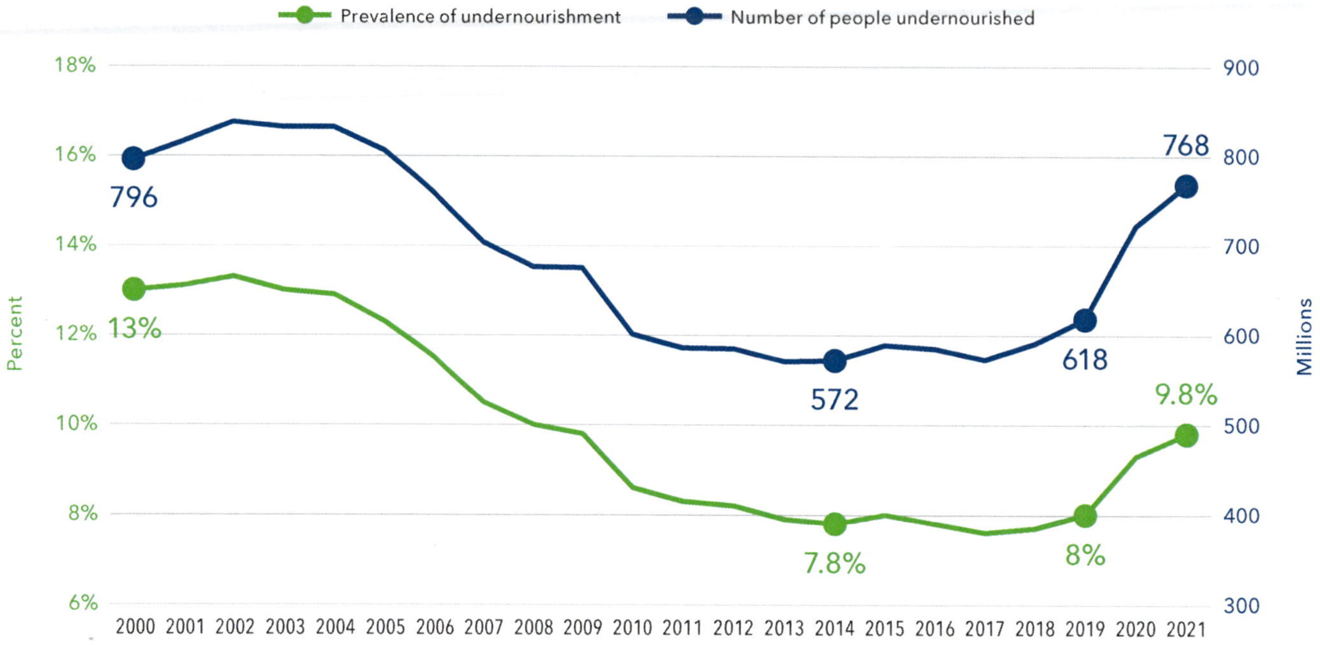

Source: FAO, IFAD, UNICEF, WFP, and WHO, *The State of Food Security and Nutrition in the World 2022* (Rome: FAO, 2022).

Note: Values for 2021 are projected; the figure shows the mid-point of the projected ranges. These figures reflect chronic hunger; see Chapter 2 on different measures of food insecurity.

that are affecting growing numbers of people. The chapter presents key recommendations from the report's thematic chapters, which explore how governments and other key stakeholders can better prepare for and respond to shocks and crises. We also consider the cornerstones of a more effective response to crises: effective governance and sufficient and flexible funding. The regional section of the report reviews how crises have impacted six major world regions in recent years, and how these developments signal new challenges and opportunities. We hope this report helps to advance a new paradigm for crisis mitigation and response, one that facilitates robust recovery and improved stability for all.

VULNERABILITY OF FOOD SYSTEMS AND FOOD SECURITY

Food systems were facing threats well before the COVID-19 pandemic. In the years before the pandemic, global development progress had started stagnating and even reversing in some places – a marked change following several decades of dramatic declines in hunger and poverty. In 2014, 572 million people were undernourished – a record low. But by 2019, this number had grown to 618 million, largely due to conflict, weather-related disasters, and economic downturns in many countries (Figure 1).[1]

During the past few years, multiple shocks have worsened this reversal in progress. The pandemic triggered a global recession, widespread labor shortages, food losses, and transport bottlenecks, which affected both the quantity and quality of available food. This likely increased the number of undernourished by 196 million people, raising the total to 768 million by 2021.[2] In 2020, an astounding 3 billion people could not afford a healthy diet.[3] This constellation of factors also set back achievement of gender equality by more than 30 years, as measured by changes in the World Economic Forum's Global Gender Gap Index between 2020 and 2022.[4]

As the recovery from COVID-19 began, prices surged for food, fuel, and fertilizer, creating new

BOX 1 FOOD AND FERTILIZER CRISIS, 2021-2022

In 2021, food prices rose to their highest levels in a decade as a result of weather shocks, strong demand associated with recovery from the COVID-19-induced recession, lingering supply chain disruptions, and record low inventories for wheat, corn, and soybeans. High natural gas and coal prices also pushed fertilizer prices to record highs. In the aftermath of the February 2022 invasion of Ukraine, food and fertilizer prices spiked even further, causing serious harm not only to wheat-importing countries, many in the Middle East and North Africa, but also to many other low- and middle-income countries. Even though many international commodity prices began to fall by mid-2022, they still remain above the historical pre-COVID-19 average, and domestic inflation remains rampant in many low-income countries.[1]

The impacts on food and nutrition security and poverty are likely to be dire. Simulations run by IFPRI researchers show that the global price shocks may have caused national poverty headcount rates to rise by as much as 7.7 percentage points and undernourishment by up to 4.4 percentage points.[2] In Egypt, for example, 48 percent of households have already reported eating less food to reduce expenses, and 75 percent have reported eating less chicken and eggs, key sources of protein.[3]

The outlook for 2023 remains critical.[4] Global stock-to-use ratios for grains remain at or below the lows of recent years. These could reach critical levels if global staple food production falls due to greatly reduced harvests in Ukraine, projected drought conditions in the Southern Hemisphere, decreased fertilizer application resulting from relatively high fertilizer prices, new weather shocks, or other shocks caused by the war in Europe or elsewhere (Figure). Moreover, many low-income countries face significant macroeconomic problems, and the share of low-income countries in debt distress has increased by 60 percent since 2015. Efforts to respond to this crisis could be improved with robust early warning systems, donor transparency and coordination, and a shift toward crisis resilience.

Global ending stocks, excluding China

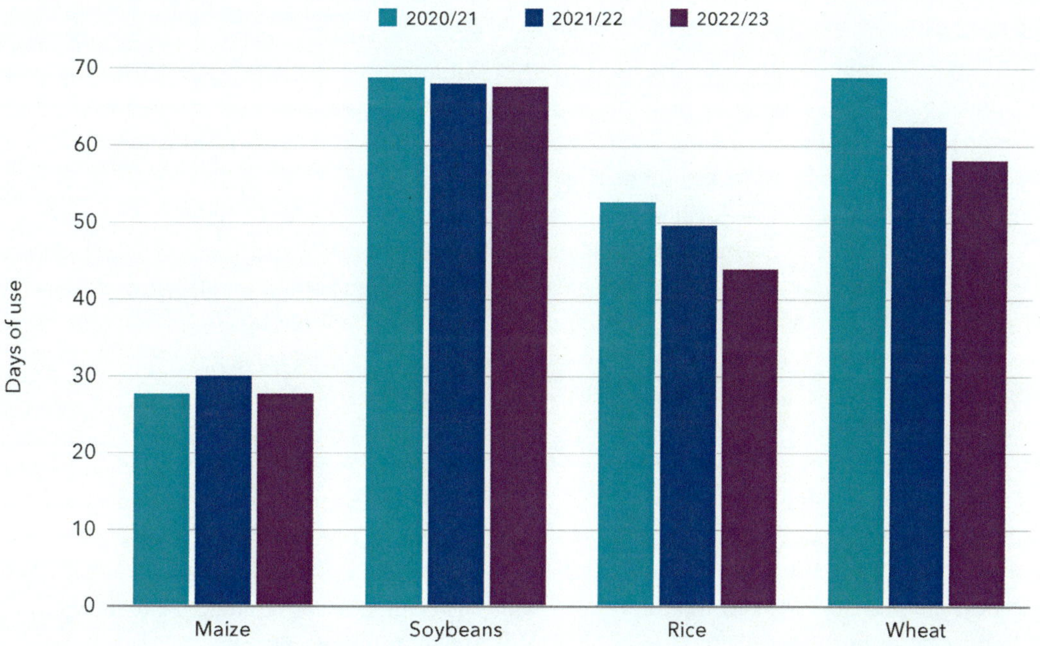

Source: Data from US Department of Agriculture, Foreign Agricultural Service, Production, Supply, and Distribution online, accessed January 2023.

FIGURE 2 Trends in extreme weather events, droughts, and floods, 1900–2022

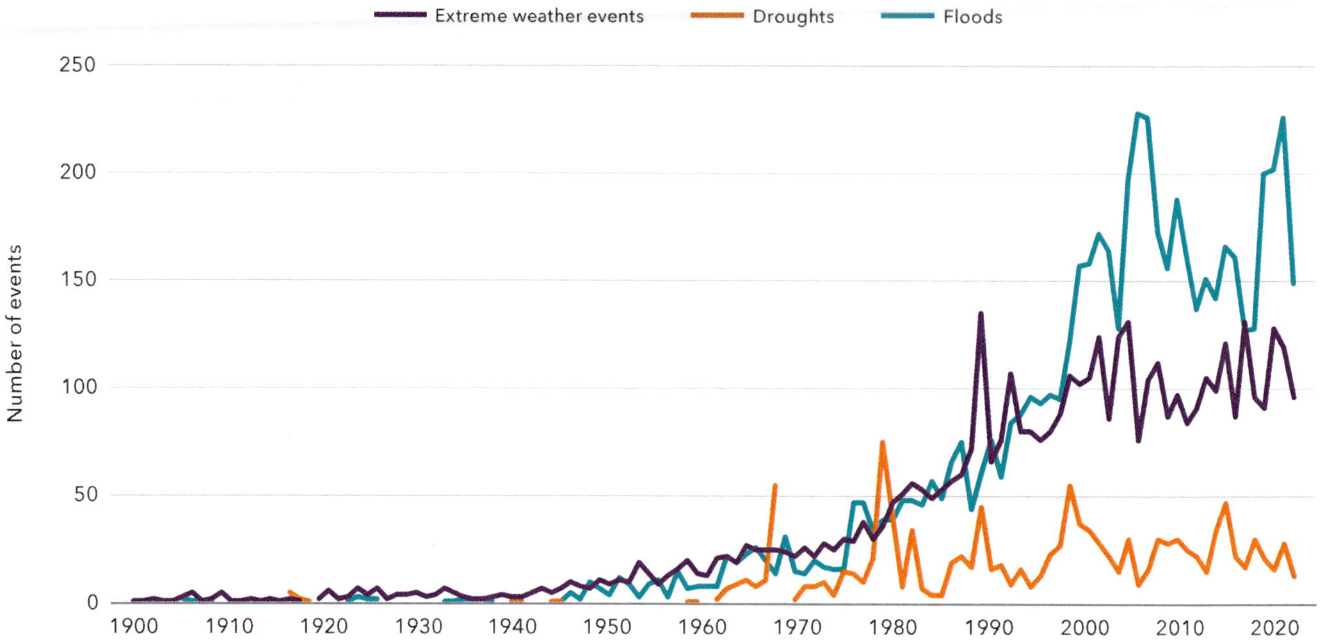

Source: Data from EM-DAT, accessed January 2023. https://emdat.be/

Note: Extreme weather includes severe storms, tornadoes, sandstorms, and extreme temperatures, among other events.

problems that were exacerbated when Russia invaded Ukraine in February 2022. International food prices subsequently rose another 32 percent,[5] and fertilizer prices tripled (Box 1).[6] Of the countries that were already in a food crisis in 2021, more than half depended on Russia and Ukraine for wheat imports, heightening risks for their populations.[7] International food and fertilizer prices have since fallen but remain high by historical standards, and many low- and middle-income countries (LMICs) are plagued by rising domestic inflation and depreciating currencies. As a result of these compounding crises, as many as 205 million people in 45 countries experienced crisis-level acute food insecurity or worse by 2022, a number that has nearly doubled since 2016.[8] Most recently, in early 2023, a severe earthquake killed tens of thousands across Syria and Turkey and left many homeless, further intensifying the level of crisis for these countries.

Shocks to food systems can take many different forms and vary dramatically in their impacts. When they lead to severe disruptions that cause a surge in acute food insecurity, these shocks are deemed a food crisis (see Chapter 2 for the technical definition of a food crisis). Whether a community, country, or region is resilient to a shock – or is at risk of a food crisis – depends on many factors. Past experiences show that crises rarely arise from isolated shocks to food systems. They are often compounded, and their negative effects intensified, by long-term sources of fragility, including poverty, climate change, gender and social inequalities, poor governance and lack of trust in public sector institutions, and lack of social cohesion.

Threats from climate change loom especially large for many countries, especially those in Africa. Climate change is rapidly intensifying, increasing pressure on food systems, rural livelihoods, and ecosystems more broadly.[9] While some places may benefit from a longer growing season amid rising temperatures, changing weather patterns and advancing desertification have reduced the average growth in agricultural productivity by as much as 21 percent since 1961. This decline in growth, which is expected to worsen, is most harmful to tropical agriculture.[10]

Climate change is also triggering more frequent and extreme weather events (Figure 2), with devastating impacts on food systems and human lives, especially in more densely populated and water-scarce regions of LMICs. In 2022, flooding in Pakistan displaced more than 33 million people, and an ongoing drought in the Horn of Africa killed 7 million livestock.[11] Climate change, along with poor agricultural practices, can increase the risk of plant diseases, pests, and zoonotic diseases. Projections from IFPRI's IMPACT model find that 65 million more people will be undernourished by 2030 and as many as 72 million more by 2050 with climate change, as compared to a scenario without climate change.[12]

Climate change also affects conflict and displacement in multiple ways.[13] In 2020, about three-quarters of internally displaced people (IDPs) were forced to relocate by disasters – mostly weather-related.[14] Conflict accounts for the other quarter, including in Somalia and Yemen, where famine warnings have recently been issued. In many places, conflict and climate change both contribute to crisis situations, most notably in Syria, Afghanistan, and South Sudan, where numbers of IDPs and refugees are high. Countries enduring conflict are particularly vulnerable to climate-induced shocks,[15] which can act as a threat multiplier that further increases insecurity, violence, and migration as resources become scarce. Recent events highlight this complex relationship: of the more than 200 million people facing acute food insecurity in 2022, most live in protracted crisis situations – that is, situations marked by prolonged civil strife and conflict, repeated weather shocks, and economic decline, or some combination thereof.[16]

UNEQUAL IMPACTS

ECONOMIC VULNERABILITY

Recent crises highlight the vast differences in how food system shocks affect the rich and the poor – both countries and their vulnerable populations. In general, LMICs have fared worse throughout many recent shocks, due to limited budgets to enact stimulus and social protection measures, reduced remittances from high-income countries, and rapidly rising import bills for food and agricultural inputs. Within these countries, vulnerable populations bear the brunt of crises. These groups – which include rural smallholders, the urban poor, the landless, IDPs, and refugees – can be made even more vulnerable by other compounding factors, such as gender, age, ethnicity, and social class.

Food system shocks are felt most severely in fragile and conflict-affected settings, where 1.5 billion people currently live. The 2021 UN Food Systems Summit (UNFSS) underscored this burden, noting, "most hungry people are in fragile and conflict-affected places…[where] it is especially difficult to transform food systems and to meet the needs of the most vulnerable and marginalized people." On average, 30 percent of people in countries facing protracted crisis situations live in extreme poverty – a situation that can prevent them from adapting to and recovering from shocks.[17]

Coping strategies can affect food and nutrition security, as well as long-term well-being. Shifting to cheaper, less nutritious staple foods, for example, is a common coping response among the poor, a practice that has increased amid pandemic-related food shortages and rising prices driven by the Russia-Ukraine war. Other damaging strategies include selling off productive assets and reducing spending on education and health – particularly for girls. Earlier marriage of girls is another response that leads to lasting harm.[18] Migration, either voluntary or forced, can have negative health implications and create challenges for livelihoods and access to productive resources, for both migrants and their host communities. However, migration can also help households escape crises, diversify risks, and expand income-generating activities.[19]

Forced migrants – including IDPs and refugees – are among the most vulnerable. By 2022, a projected 103 million people were forcibly displaced worldwide (Figure 3). Of this group, 80 percent experienced acute food insecurity and high levels of malnutrition.[20] Russia's invasion of Ukraine has triggered Europe's largest refugee crisis since World War II, with nearly 8 million people fleeing the war. Despite this, LMICs host 83 percent of the world's international refugees, many of whom have been displaced for years and even decades.[21]

FIGURE 3 Forcibly displaced people worldwide

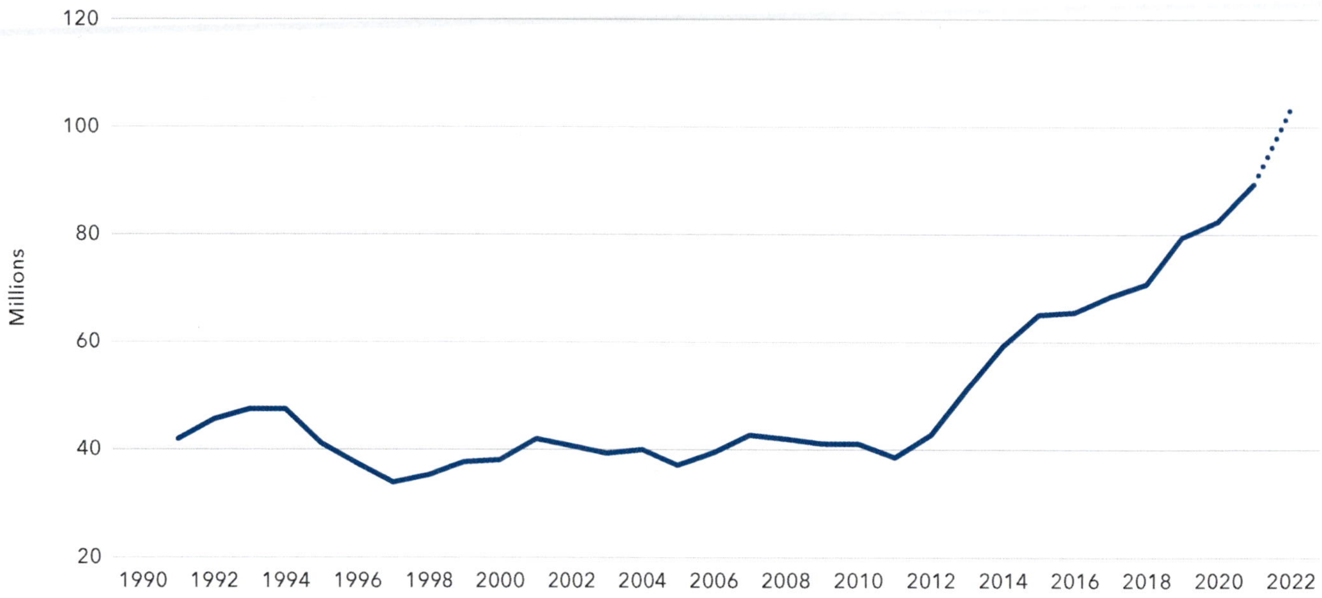

Source: UNHCR, Refugee Data Finder, updated October 2022. https://www.unhcr.org/refugee-statistics/

Note: Includes internally displaced people as of end-2021, refugees as of mid-2022, asylum-seekers as of mid-2022, and other people in need of international protection as of mid-2022.

The number of IDPs is almost double that of international refugees, with about half living in Syria, Colombia, the Democratic Republic of the Congo, and Yemen.

GENDER AT THE CENTER OF FOOD CRISES

Women are disproportionately harmed by crises, given the structural and normative barriers that limit their resilience and ability to respond effectively. More so than for men, shocks reduce women's access to food and dietary diversity, decision-making power within their households, assets, services like healthcare, and physical safety, and also deepen their time poverty.[22] These vulnerabilities stem from women's already limited access to resources, technologies, and services – which is intensified by shocks and crises – as well as to channels of power and influence that could help them benefit from crisis response policies and programming.[23]

Rural women in LMICs face barriers not only to accessing land, water, and other productive resources,[24] but, just as importantly, to accessing and benefiting from complementary resources, technologies, and services needed for agricultural production and participation in the food system.[25] For example, having less social capital can limit women's access to technology (such as modern agricultural inputs, mechanization, labor-saving technologies, and information and communications technology [ICT]), agricultural extension and advisory services, and financial services (credit, formal savings, and insurance). Crises can intensify these gender gaps – as resources become increasingly scarce, women's access is likely to decline further. Shocks can also intensify the burden of unpaid care work for women, such as providing food, collecting water, and caring for the sick, and increase gender-based violence.

Shocks and crises can also disrupt critical social protection structures and support. For example, extreme weather events or a pandemic like COVID-19 can prevent women from accessing government identification cards needed for relief programs, or make it difficult to collect payments. In times of crises, governance structures may also prove more dysfunctional or reduce funds for social protection.

Migration further complicates gender issues. Women and girls account for about 50 percent of IDPs and refugees, but in some places they make up a much larger share. In addition, children account for more than 40 percent of all displaced people.[26] However, when women remain at origin and men migrate – as often occurs with economic shocks – women may shift from contributing as family workers to become primary farmers.[27] Without access to key resources or greater decision-making power, this increase in responsibilities and workload can leave women worse off.[28]

A WAY FORWARD: BUILDING ON WHAT WORKS

Although the rise in food insecurity and poverty is alarming, food systems showed major strengths during recent crises. Understanding these strengths can help stakeholders rethink the way forward and build on successes as they respond to new crises.

In recent decades, a range of transformational developments has increased the resilience of food systems. Trade has helped countries to secure alternative suppliers during supply shocks, though export restrictions during crises can still pose a threat.[29] Urbanization and rising incomes in LMICs have sparked demand for more diverse foods, including animal-source foods and fruits and vegetables. In response, value chains have expanded and diversified, potentially improving the ability to meet food and nutritional needs in the face of shocks, while creating new livelihood opportunities. Value chains also provide inputs and services to rural producers, which can increase resilience in the agriculture sector. In rural and urban areas, social safety nets have been more widely adopted, providing food security and better economic opportunities for women and men. In many places, the growing empowerment of women has strengthened their decision-making role in food systems, helping them to derive greater benefits from these systems.

In addition, efforts have been expanded to predict crises and proactively reduce their impacts through programming and effective governance and institutions. Several new approaches have been tested around the world, including anticipatory action programs, forecast-based financing, and the scaling-up of innovative social safety nets. The upward trend in migration has, when managed well, expanded job opportunities (particularly for youth) and helped households support their livelihoods, make investments, and build resilience.[30] Taken together, these developments warrant policies that capitalize on their capacity to support resilience.

A NEW, MORE PERMANENT RESPONSE

As the world reflects on lessons learned from recent food system shocks, now is an opportune moment to rethink our approach to food crisis response by building on existing innovations and exploring new solutions. Traditional crisis response has focused on humanitarian and emergency food aid, but a more systematic and sustainable approach is needed to address protracted crises, which are likely rising.[31] Research tools are already available to the international community and national governments to help them not only predict, monitor, and respond to crises, but also to govern for resilience and equity. Shifting toward longer-term and more permanent "crisis resilience" is critical.

The thematic chapters in this report explore some of the promising policies, programming, and tools for developing a strong response to increasing and intensifying shocks. These can help us better predict and prepare for crises, address crises when they occur, and build more resilient and equitable food systems.

PREPARING FOR CRISES

Early warning systems, especially in combination with anticipatory action efforts, can facilitate both immediate humanitarian responses and the integration of aid with longer-term development strategies. Existing systems must be improved to better address the growing complexity of crises, including climate-related events and conflict situations.

Early-warning, early-action (EWEA) systems alert policymakers and international humanitarian agencies to sudden and significant increases in acute food insecurity, signaling food crises, and provide guidance on where and when to target humanitarian efforts. A timely and effective response depends on accurately identifying and tracking different food crisis situations; understanding how they affect different populations, sectors, and places; and addressing the pressures exerted on people and food systems.

Multiple systems are already monitoring chronic and acute food insecurity as well as trends in agrifood markets, such as sudden changes in the prices of international agricultural commodities and fertilizers. This information is extremely useful, but ideally it should be consolidated and improved to shape responses more precisely. This will require filling gaps in monitoring and analysis, particularly to understand and track the drivers of crises in diverse contexts, including compound crises. It will also require better integration of existing systems to ensure that policymakers and others receive clear, timely warning signals of potential crises and guidance on priority setting. Finally, new processes are needed that allow for faster classification and response to crises, especially to identify famine, where immediate response is most crucial. Chapter 2 considers the role of early warning systems in crisis response and suggests ways to assist policymakers with defining and prioritizing responses.

The vast majority of humanitarian response is activated after a crisis occurs, delivering life-saving aid but at relatively high costs. During crises, rapid response is critical to reach households before they deplete savings or engage in damaging coping strategies, and before widespread repercussions occur, such as increased fragility. Anticipatory action frameworks help prepare and organize humanitarian aid before crises strike by allocating funds, responsibilities, and supplies in advance. These frameworks, along with innovative forms of humanitarian assistance, show promise for mitigating crises at lower costs and supporting longer-term development efforts. Once triggered by an early warning system, the anticipatory action plan can be implemented smoothly and without lengthy delays.

Anticipatory action requires monitoring data that illuminate risks, exposure, and vulnerability; information services that can reach vulnerable people and advise them on how to respond; and a clear decision support system, especially in fragile settings where government authority may be weak. Its effective delivery also depends on robust governance arrangements, which can ensure appropriate targeting and deployment. When more broadly conceived, anticipatory action can help shift the focus of crisis response toward longer-term resilience and development by incorporating nutrition-sensitive programming, making use of local procurement, and supporting local institutions and more permanent safety nets. This approach could play a crucial role in mitigating food system shocks, but currently makes up only a small percentage of humanitarian aid.

To increase adoption of these programs, more data and research are needed on the effectiveness of different humanitarian assistance approaches and anticipatory action programs for protecting food and nutrition security – particularly in fragile and conflict-affected settings. Chapter 3 discusses the potential of anticipatory action and innovative types of humanitarian assistance, how these can align with development strategies, and how further data collection and analysis can support them.

CREATING RESILIENT FOOD SYSTEMS

Social protection systems, including safety net programs that provide food or cash transfers, can both build resilience prior to a crisis and facilitate crisis recovery. They are most effective when they are flexible, shock-responsive, and well targeted. Before a crisis, safety nets help households and communities build assets, increase productive investments, and diversify income sources. During crises, social safety nets can prevent negative coping strategies that pose risks to long-term health and livelihoods. Many LMICs have dramatically expanded their social safety nets in recent years, but as the COVID-19 pandemic and recent food price spikes showed, coverage is low in the poorest countries, and many cannot access these safety nets – particularly the urban poor.

A proactive approach is needed to develop social protection systems that are highly adaptive, flexible, and inclusive, and can be quickly expanded when crises strike. Support can be scaled up more quickly and effectively by integrating these "shock-responsive" social protection systems with EWEA systems and humanitarian aid, and creating unified and digitized targeting systems. In addition, integrating social protection with gender and climate goals can further empower women and promote environmental sustainability. Given the great need to expand safety net programs, new ways to cover costs should be explored, such as integration with green financing schemes, as well as ways to reduce implementation costs, including cash transfers and mobile payments. Chapter 5 considers the role of social protection in both resilience building and crisis response, exploring how these programs have evolved over time and how best to ensure their longevity by examining financial realities, new modalities, and a greater focus on inclusion.

The successful functioning of food systems relies on agrifood value chains, including the production, processing, transport, and marketing of food. These value chains differ greatly in their structure and local contexts, which in turn affects the impact of shocks and value chain responses. Given these differences, crisis responses are likely to be more effective when tailored to the type of shock, the particular context and value chain, and if possible, the size of the affected enterprises.

The experience of the COVID-19 pandemic highlights the importance of flexibility for all types of value chains and their actors. Almost everywhere, food-related businesses that were able to digitize and develop new marketing mechanisms amid pandemic-related restrictions proved hardier that those that were not.[32] Private sector actors can increase their businesses' resilience by investing in improved and innovative tools, such as climate-smart agriculture and new forms of insurance. Governments can provide support by creating a regulatory and business environment that fosters value chain innovations and ensures that women-owned enterprises can take advantage of them. Governments can also support an open trade policy to facilitate the diversification of value chains. Before and during crises, government monitoring can help to ensure the continuation of private trading and guide it where needed.

Chapter 4 explores the strengths and vulnerabilities of value chains, with a close look at the differences in how crises affect various actors, including small and large enterprises and those owned by women and men. It shows how the capacity to innovate and policies that allow trade and innovation to continue are critical to both quick recovery and long-term resilience.

SUPPORTING AND EMPOWERING THE MOST VULNERABLE

Building resilience among the most vulnerable populations, particularly women and forced migrants, can reduce the impact of crises when they occur and speed recovery. Food system resilience must therefore include a strong focus on enhancing livelihoods and inclusion. These efforts must prioritize the needs of the most vulnerable in the short term, ensuring access to food and vital services, but also build their resilience and capacity for the longer term.

Empowering women amid crisis situations is particularly important, given that they shoulder a disproportionate share of negative impacts and often deplete their assets or compromise their diets as a coping mechanism. A first step to increase equity involves improving the quality of gender-disaggregated data collected before and during crisis situations, including on women's access to programs meant to support them. Innovative methods, such as phone surveys, can facilitate data collection in fragile and conflict-affected settings. When decision-makers have more specific information about the different women who are enduring various negative effects, policies and programming can be tailored to better support them. Effective policy responses along with legal protections will also need to account for the barriers that women face to participating in food systems, their domestic work burdens, and the likelihood of gender-based violence, all of which are likely to increase amid crises.

Efforts must also be made to increase women's political participation and amplify their voice and agency in their communities. In particular, women's voices must be included in peace processes and high-level positions where policymaking and programming decisions are made, so that crisis responses improve rather than erode gender equality. Such policy responses can empower and create opportunities for women while also addressing the adverse impacts of crises. Finally, supporting women's access to resources and technologies, including mobile phones, can help them better weather crises.

Being explicit about gender targets and tracking progress is central to promoting gender equality amid crises. For the long term, effective gender-focused interventions including cash transfers, self-help groups and other civil society organizations, and/or technical and vocational training, among others, can help women in diverse settings build resilience to shocks and crises. Chapter 6 explores what we know about the gendered impacts of crises, reviews the most important data gaps, and provides recommendations for ensuring that crisis responses address inequities.

Conflict and climatic and economic crises often trigger forced migration (Chapter 7), creating challenges and opportunities for migrants and their sending and host communities. Although people forced to migrate often face high risks and food insecurity, migration can play an important role in improving individual livelihoods and economic development. Forced migrants and refugees have been shown to make positive contributions to their host communities' economies, and remittances to sending communities can provide substantial benefits as well.[33] Thus, all stand to benefit from policies that facilitate economic and social integration, including cash transfers, training programs, and the right to work and choose a place of residence. However, forced migration can strain host communities when resources and opportunities are limited, requiring efforts to limit migration from sending communities while strengthening the absorptive capacity of host communities.

Governments, NGOs, and development organizations can better address the root causes of forced migration through innovative data collection and research, especially on irregular migration and the needs of women. They can build the capacity of host communities by investing in infrastructure and services and designing policies that expand the benefits of migration and limit harms. Innovative approaches hold great potential to accelerate the transition from humanitarian action to longer-term development, such as by aligning social protection and climate action objectives to mutually support peace, security, and sustainability. Attention must also be paid to those who remain behind, because they often lack the resources or social networks needed for migration, and are least capable of recovering from a crisis. Chapter 7 reviews key facts about forced migration and provides recommendations to ensure that policies increase the benefits of migration and reduce detrimental impacts on migrants, host communities, and sending communities.

FOUNDATIONS FOR BETTER CRISIS RESPONSE

Improving international and national responses to food crises cannot be done without accountable governance and effective institutions, policies, and programming, as well as reliable funding and oversight to ensure that responses address immediate needs and long-term resilience.

GOVERNANCE

Effective governance at all levels is critical to advancing early warning, anticipatory action, and policy responses that are sustainable and responsive to the compounding drivers of crisis. Institutions and public sector incentives must support government accountability (that is, responsiveness to citizens' needs and preferences), as well as the equitable, reliable, and cost-effective provision of infrastructure and services. This requires making the best use of government investments (rather than wasting or squandering them), and ensuring the effective deployment, communication, and continuity of anticipatory action, humanitarian assistance, social protection, and other programs that are critical to averting and addressing shocks and crises.[34] Effective governance can also minimize market disruptions and incentivize private sector investments to promote resilience. Finally, it can more broadly contribute to trust and social cohesion to help avoid internal conflicts and future crises.[35] The pillars of the UN's far-reaching Sendai Framework for Disaster Risk Reduction integrate good governance structures, and many measures of crisis preparedness include some version of governance, whether viewed as the provision of planning services or effective communication between leaders and their citizens.[36] In many instances, good governance mechanisms have been shown to improve disaster preparedness.[37]

Many promising approaches exist to build effective governance. For example, transparency and the free flow of information, including through ICT that connects government with citizens, can help make governments more accountable.[38] Improving the incentive environment for bureaucrats and frontline service providers can ensure that they are hired and promoted for delivering what matters to citizens. Education, training, and transparent policymaking can help guarantee that the voices of women and other vulnerable groups are included in crisis responses to broadly support gender equality and social inclusion. To hold governments accountable, international and local actors can use research tools to track social, economic, and environmental risks and to monitor and evaluate policy responses to crises.

FINANCING MECHANISMS FOR CRISIS PREVENTION, PREPAREDNESS, AND RESPONSE

The developments of the past few years have dramatically increased the need for better crisis response funding. In 2023, the UN Office for the Coordination of Humanitarian Affairs appealed for US$52 billion in funding for humanitarian assistance and protection, a 461 percent increase since 2012. Funding received in 2022 amounted to $24 billion, or only 47 percent of the need.[39] Governments were forced to spend record amounts on social protection in response to compound crises, even as programs faced disruptions due to these very shocks. In 2022, 170 economies announced, implemented, or enhanced more than 1,000 social protection and associated programs to mitigate the impacts of inflation, a fourfold increase from April 2022 to December 2022. About $711 billion, equivalent to 0.7 percent of global GDP, was invested in social protection in 2022.[40]

This funding must be increased to meet growing needs. Although some crisis funding increased in 2022 – such as the International Monetary Fund's (IMF's) opening of a temporary food shock window to quickly channel funds to countries impacted by the global food crisis – far more is needed, especially for crisis preparedness, resilience building, and support for humanitarian-development-peace approaches. Smart investments to build resilient food systems, while costly, are far more cost-efficient and effective than reacting to crises after they occur.

The finance lever of the UNFSS estimates that it would cost between US$300 billion and $400 billion per year through 2030 to transform food systems for sustainability and resilience.[41] Some of this investment can be used to expand credit market access to smallholders and small and medium enterprises in LMICs. Credit can provide these businesses with a short-term financial cushion during shocks and an opportunity for long-term investment in resilience-enhancing technology and practices. For example, producers can use credit to invest in solar power, cold storage, or drought-resistant crop varieties that will help address climate threats. At the national and international levels, financial flows should be redirected toward more crisis-resilient technology, practices, and infrastructure. Forecast-based finance schemes, currently being implemented by some agencies, could be expanded and deployed in fragile settings for beneficiaries and locations that have been identified ahead of time.

A key strategy to redirect these funds involves repurposing the more than $600 billion in global spending that goes for agricultural support. Currently, much of this financing supports activities that are inefficient and unsustainable. Some funds could be reallocated to incentivize the adoption of more sustainable practices such as no-till farming, and invested in agricultural research and development aimed at traditional targets such as productivity gains, as well as new targets such as improved resilience.[42]

Policymakers can also do more to shift private investment toward crisis prevention and resilience, given that private sector investment in food systems far outweighs that of governments. Both the quantity and quality of private sector funding for resilience can be improved by creating an enabling environment for private sector actors to invest, and incentivizing investments that support livelihoods and sustainability. Business opportunities to implement Sustainable Development Goal actions related to food and agriculture could garner $2.3 trillion annually for the private sector by 2030, while requiring an annual investment of only $320 billion.[43] Conversely, enacting rules for private investors, such as requiring publicly traded companies to disclose environmental and climate-related risks, could more closely align financial incentives with the SDGs and the Paris Climate Agreement.[44] Development banks could also use their funds to de-risk and crowd-in private investment through blended finance or food systems bonds. For example, the Bridgetown Agenda, promoted at the recent climate COP27, called for $500 billion in IMF Special Drawing Rights to be used to attract private investment in resilience for low-income countries at the frontlines of the climate crisis.[45] Ultimately, all such changes to current financial flows would prevent even greater future costs in the form of crisis response, economic disruption, and loss of life.

CONCLUSION

The first years of this decade exposed the many vulnerabilities of our food systems, which employ 2 billion people and sustain and nourish all of the world's 8 billion people.[46] Food systems are not only susceptible to increasingly complex and compounding shocks, but are also closely intertwined with other essential systems – climate and environmental services, trade and the economy, infrastructure, governance, healthcare, and social protection. Failures within these systems can cause crises in our food systems, and in turn, weaknesses in our food systems can drive environmental degradation, conflict, economic disruptions, and poverty and inequity.

Using food systems to build a more proactive response to disaster – one that is anticipatory, flexible, and inclusive – can produce multiple benefits for food and nutrition security, poverty, livelihoods, equality, and political stability. The process of building and improving crisis responses should be rooted in high-quality evidence: robust data, state-of-the-art tools, and policy analyses and scenarios developed by research organizations and networks like IFPRI and CGIAR. This evidence can help policymakers, donors, the international development community, and the private sector to move quickly in times of need. Increasing crises in human systems and the natural world will not abate in coming years – the time to step up our efforts to develop a more permanent, sustainable response is now.

Now is an opportune moment to rethink our approach to food crisis responses by building on existing innovations and exploring new solutions.

CHAPTER 2

Food Crisis Risk Monitoring

Early Warning for Early Action

ROB VOS, ARIF HUSAIN, FRIEDERIKE GREB, PETER LÄDERACH, AND BRENDAN RICE

Rob Vos is director, Markets, Trade, and Institutions Unit, International Food Policy Research Institute (IFPRI). **Arif Husain** is chief economist and director of Research, Assessment and Monitoring, World Food Programme. **Friederike Greb** is an economist, Research, Assessment and Monitoring, World Food Programme. **Peter Läderach** is CGIAR co-lead for climate security, The Alliance of Bioversity and CIAT. **Brendan Rice** is a research analyst, Markets, Trade, and Institutions Unit, IFPRI.

KEY MESSAGES

- Early-warning, early-action systems provide alerts of potential food crises – identified as sudden and substantial increases in acute food insecurity – as well as guidance to policymakers and international development agencies about needs for humanitarian action.
- Use of different methodologies and varying coverage of vulnerable populations mean different early warning systems for acute food insecurity can yield dissimilar estimates of the severity of food crises.
- Local food security monitoring systems are poorly connected to systems that track global food and agriculture market trends. Monitoring of acute food insecurity and chronic food insecurity are poorly integrated at the country level. This leads to differing interpretations of the nature and magnitude of food crises.
- Existing systems pay insufficient attention to structural vulnerabilities that determine how different shocks, including global price shocks, affect food insecurity in particular contexts and compound other causes of acute food insecurity, such as poverty, conflict, and climate change.
- Famines are the catastrophic expression of severe food crises. Today's famine-like contexts are mostly driven by conflict. Conflict typically impedes the data collection required by existing protocols for declaring famine, which can delay humanitarian action, at the expense of a preventable human toll.

To increase the effectiveness of early warning systems, it is important to:

- Expand the country coverage and frequency of consensus-based acute food insecurity analysis.
- Revise the protocol for declaration of a famine to ensure it is operational in conflict-affected locations.
- Better integrate the various types of early warning systems for food crises through much stronger collaborative efforts across responsible international organizations, with support from the research community and in consultation with policymakers, development agencies, and local actors.
- Improve monitoring of risk factors and structural causes of crises to support the development of real-time early warning systems that are able to anticipate and potentially help prevent food crises through timely and well-targeted responses.
- Strengthen analysis of factors driving crises in particular places – including global supply and price shocks, how these are transmitted to local contexts, what structural vulnerabilities increase or mitigate their impact, and how they affect acute and chronic food insecurity – to inform long-term responses that build resilience and reduce the risk of food crises.

Global and national agrifood systems are vulnerable to a variety of shocks that have caused major disruptions to food production, markets, and livelihoods over the past two decades, and have set back efforts to reduce poverty, food insecurity, and malnutrition. Currently, the world is contending with the global repercussions of the Russia-Ukraine war. In many countries, the impact of the war is compounded by local conflict, weather shocks, lingering effects of COVID-19, macroeconomic instability, and weak coping capacity. These concurrent crises have led to a sharp rise in both acute and chronic food insecurity since 2017, especially in developing countries. According to estimates from the United Nations agencies, chronic food insecurity – measured as the number of people with prolonged insufficient food energy intake – rose from around 573 million in 2017 to as many as 828 million in 2021 (Figure 1A). Acute food insecurity – measured as food deficiency affecting lives at any given point in time – almost doubled between 2016 and 2022, from 108 million people in 2016 to 205 million in 2022 in 45 food crisis countries (Figure 1B). Estimates of the World Food Programme (WFP), which considers more countries, suggest that as many as 349 million people in 79 countries faced acute food insecurity in 2022.[1]

This rapid rise in food insecurity has placed tremendous pressure on governments and humanitarian and development partners to respond, despite limited financial resources. These actors are also hindered by insufficient information needed for prioritizing policies, investments, and other interventions and for balancing responses to immediate impacts with investments in longer-term resilience. Governments and international agencies increasingly need more effective early warning systems that provide timely and accurate projections to inform policies for immediate and longer-term responses.

Multiple early warning systems exist to monitor food crisis risks. Among those that directly monitor acute food insecurity, there is considerable overlap and sometimes seemingly conflicting information because of differences in methods, population coverage, and frequency of data collection. Other warning systems focus on global food and agricultural market trends, providing useful information

FIGURE 1 Chronic hunger and acute food insecurity

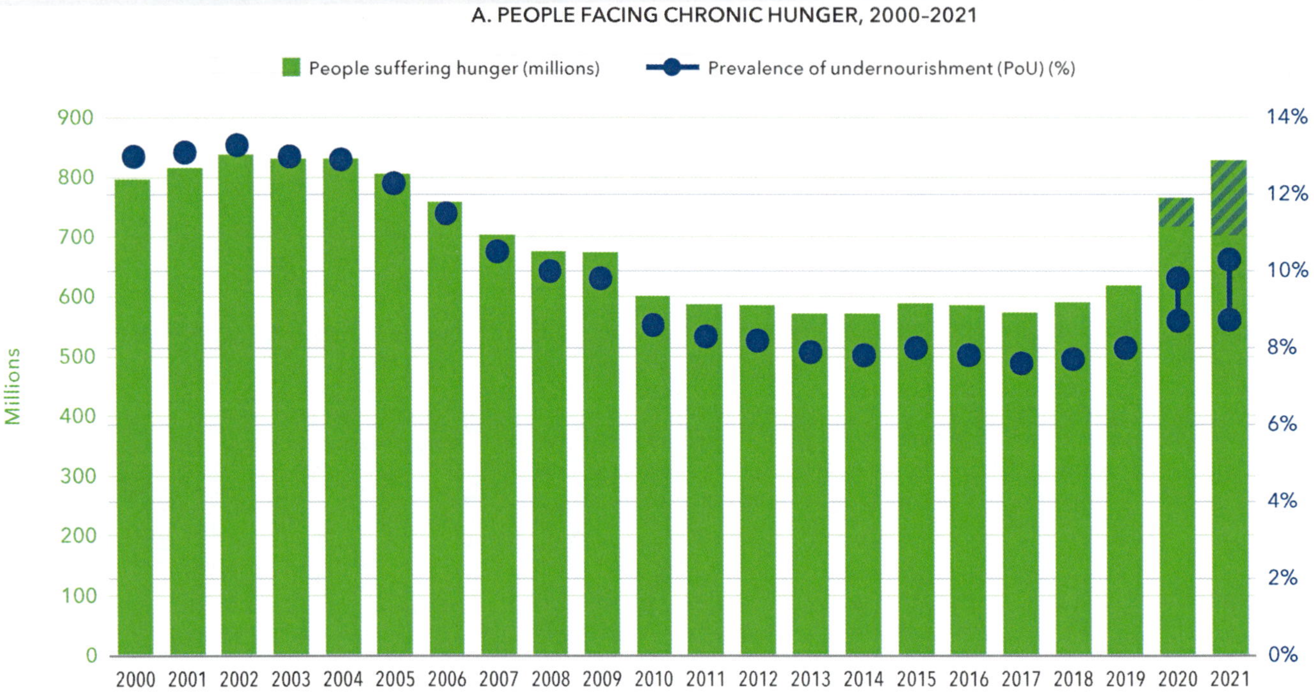

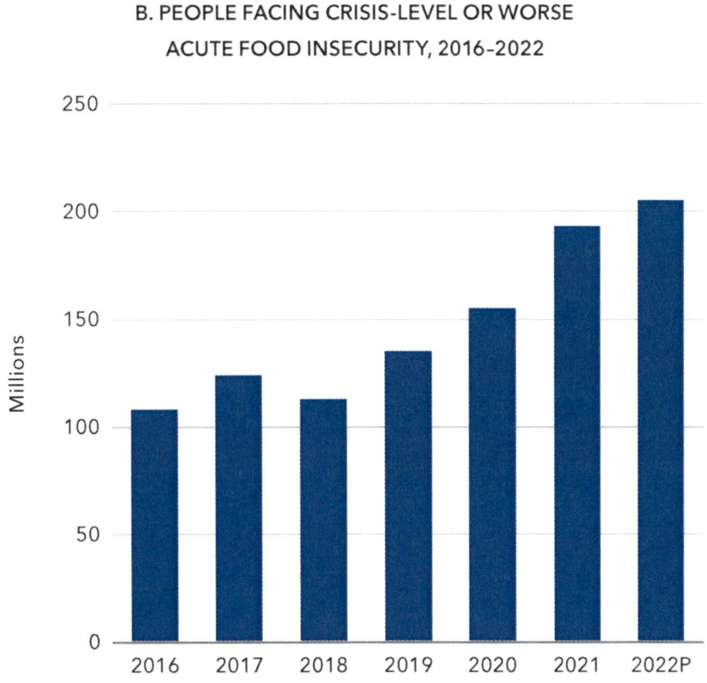

Source: For Figure 1A, FAO, IFAD, UNICEF, WFP, and WHO, *The State of Food Security and Nutrition in the World 2022* (Rome: FAO, 2022); For Figure 1B, FSIN and GNAFC, *2022 Global Report on Food Crises: Joint Analysis for Better Decisions* (Rome: 2022); and FSIN and GNAFC, *2022 Global Report on Food Crises: Mid-Year Update* (Rome: 2022).

Note: In Figure 1A, numbers for 2020 and 2021 indicate the projected range. In Figure 1B, number for 2022 is projected.

BOX 1 DEFINING FOOD INSECURITY

Food insecurity is broadly defined as the lack of secure access to sufficient safe and nutritious food needed for normal human growth and development and for an active and healthy life.[1]

Chronic food insecurity describes a situation where people are unable to meet their minimum food requirements (usually defined as minimum intake of calories) over a sustained period of time – usually over the course of a year or longer. Chronic food deprivation is most closely associated with "hunger," that is, the prevalence of undernourishment, as monitored by the Food and Agriculture Organization of the United Nations (FAO) and other international organizations.[2]

Acute food insecurity is defined as any manifestation of food insecurity at a specific point in time of a severity that threatens lives, livelihoods, or both, regardless of the causes, context, or duration.[3] Acute food insecurity is highly susceptible to change and can manifest in a population within a short amount of time, as a result of sudden changes or shocks that affect determinants of food insecurity and malnutrition.[4] Acute food insecurity can be **transitory**, in the sense that it reflects a short-term or temporary inability to meet food consumption requirements related to sporadic crises, which suggests a capacity to recover. However, situations of severe acute food insecurity often emerge in contexts where widespread chronic food insecurity already exists and where affected people have little to no capacity to recover without assistance.

about risks to (global) food availability and affordability, but do not directly link to national and localized food security risks. Better integration of these types of early warning systems and expanded capacity for data collection and analysis of complex drivers of food insecurity are needed to inform policies and rapid action to reduce the impact of, or even prevent, food crises.

This chapter provides an overview of the main early-warning, early-action (EWEA) systems now in place for identifying food crisis risks and informing responses. We describe what works, as well as the shortcomings of present systems. The chapter then discusses how better integration of currently disjointed food crisis monitoring and analysis mechanisms could create a more effective, real-time monitoring mechanism for identifying and understanding global and national threats to food security. This would allow not only swift palliative action but – importantly – could also inform the design of preventative and preemptive responses that create resilient food systems and livelihoods and reduce food crisis risks. As such, it would overcome the costly drawbacks of traditional approaches that limit early action to humanitarian assistance, which saves lives but does not address the structural vulnerabilities that may contribute to the recurrence of food crises and to the protracted nature of many food crisis situations.

EARLY WARNING SYSTEMS

Food security is a growing global concern demanding policy solutions. A "food crisis" is generally identified when rates of acute food insecurity surge at the local, national, or global level (Box 1).[2] Of the 45 countries and territories covered in the most recent *Global Report on Food Crises*, 10 reported that the number of people facing acute food insecurity increased by more than 50 percent during 2022, owing to escalating food prices, weather extremes, and conflict or insecurity (Figure 1B).[3] For example, at present, a compound crisis is unfolding in the Horn of Africa, where an unprecedented multiseason drought that began in late 2020, combined with conflict, displacement, and macroeconomic shocks, has put the region on the brink of famine.

Increases in global food prices can be an important driver of food insecurity, though the impacts are mediated by local conditions and vulnerabilities. Food prices surged in 2021 as markets faced supply bottlenecks during the COVID-19 recovery and spiked further in the first half of 2022 as a consequence of the Russia-Ukraine war.[4] Countries already facing protracted food crises before the pandemic and the war have been hardest hit by the recent surge in food prices (see Chapter 1, Box 1). Populations in all 45 crisis contexts saw the cost of a

FIGURE 2 Share of analyzed populations in crisis-level or worse acute food insecurity in 45 countries/territories, 2022

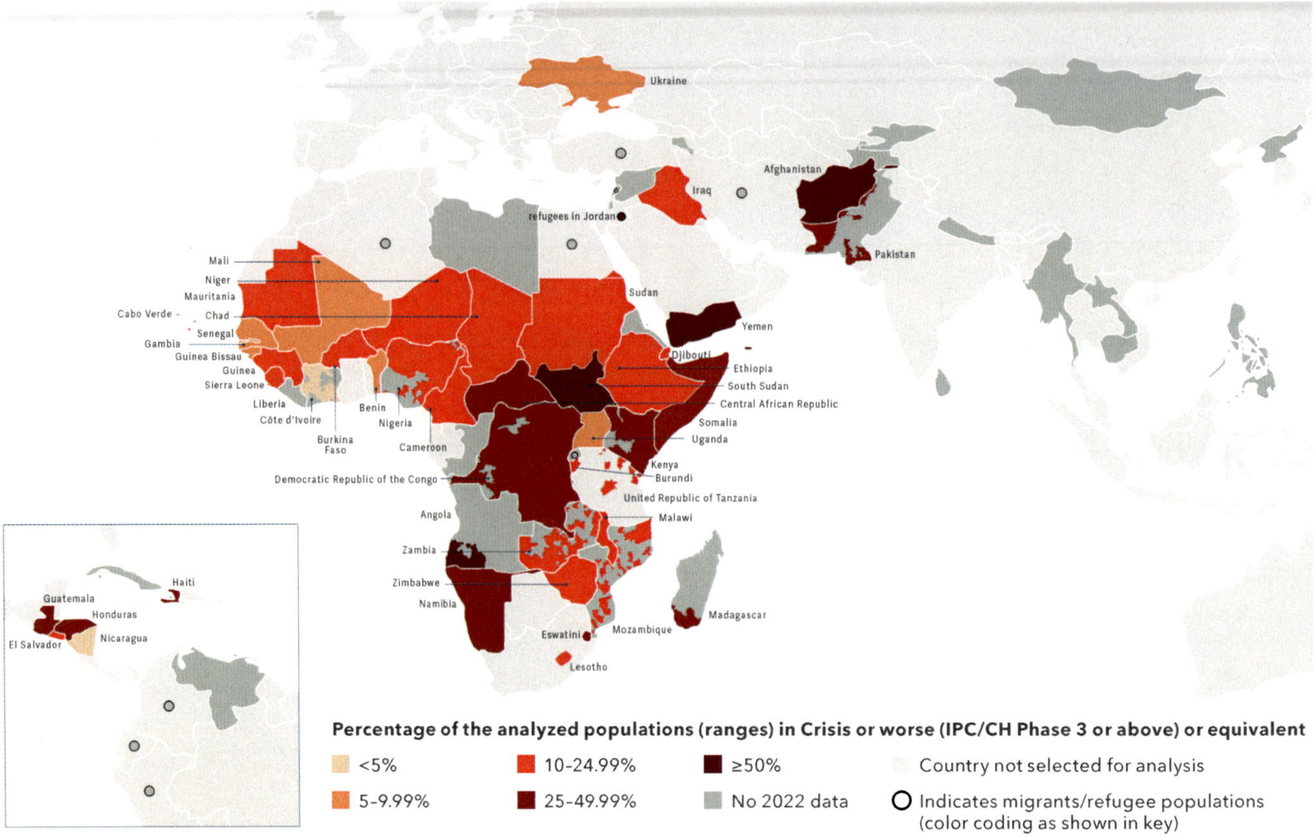

Source: Reproduced from FSIN and GNAFC, *2022 Global Report on Food Crises: Mid-Year Update* (Rome: 2022).

basic food basket increase by at least 10 percent by April–June 2022 (up from the five-year average), but people in Ethiopia, Haiti, Sierra Leone, South Sudan, Sudan, Syria, Yemen, and Zimbabwe faced annualized food cost increases of more than 75 percent.[5] Many are also suffering high general price inflation (driven by the cost of energy and other basic needs), further eroding their purchasing power.

Most food crisis countries are highly dependent on food imports and have little capacity to insulate their populations from imported food inflation, given their low foreign currency reserves, high public debt burdens, and/or depreciating national currency. International support to address this constraint (such as through additional aid, debt relief, or improved access to contingency financing) is often overlooked as a necessary companion to food assistance. Even if this support comes, it typically arrives late and is inadequate, as we saw during the 2022 food crisis – when the IMF opened its Food Shock Window only after food prices had been falling for more than four months. Consequently, governments of food crisis countries have little scope to expand social protection or other support to vulnerable populations. Protracted civil strife and weather shocks have compounded economic woes in many of these countries, such that the total number of people facing acute food insecurity at crisis level or worse (Box 2) in 45 food crisis countries increased from 155 million in 2020 to 205 million by mid-2022.[6] In Afghanistan, South Sudan, Somalia, Yemen, and parts of the Democratic Republic of the Congo (DRC), more than half of the analyzed populations at risk are considered to face acute food insecurity or worse, and famine warnings have been issued for Somalia and Yemen (Figure 2).

We focus here on two main types of early warning systems: *acute food insecurity early warning*

FIGURE 3 Agricultural market and food insecurity early warning systems by type and frequency of information

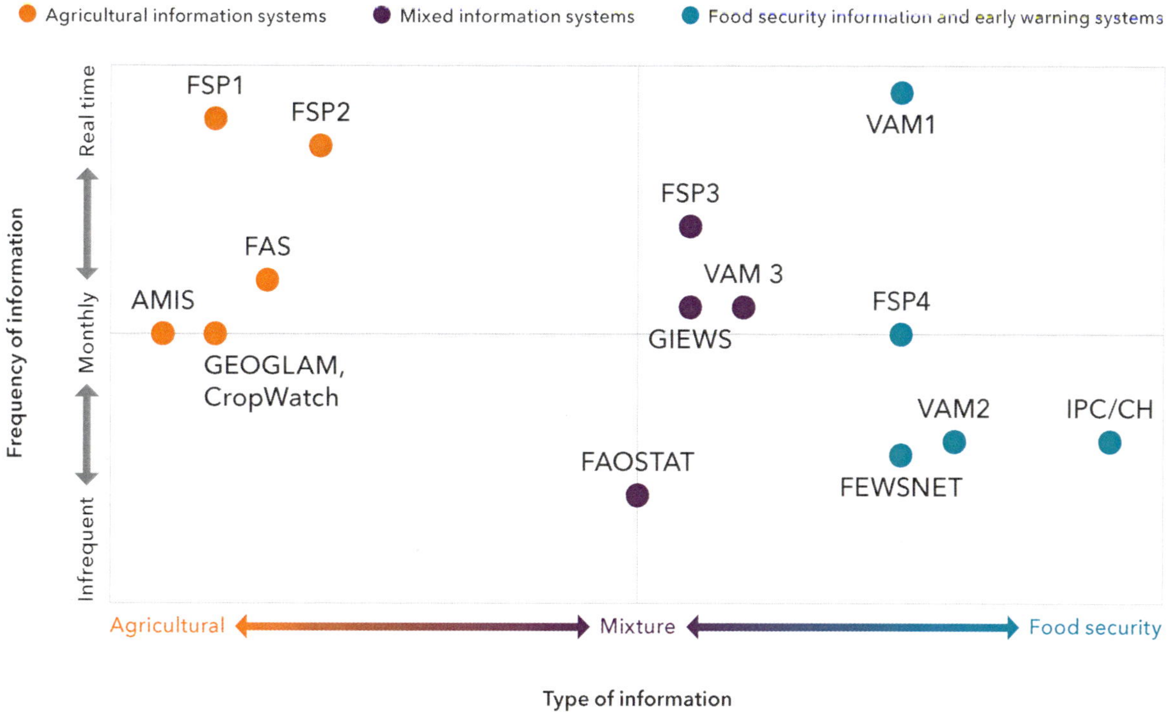

Source: Table A.1 (end of this chapter) and N. Haan, M. Van Dijk, and W. Rossi Cervi, *Food Security and Agriculture Information Systems Landscape Analysis* (London: CASA and UK Aid for the Foreign Commonwealth & Development Office, 2021).

Note: FSP1, FSP2, FSP3, and FSP4 refer to, respectively, IFPRI's Food Security Portal's (1) commodity price and volatility monitoring system; (2) trade and fertilizer restrictions trackers; (3) domestic food price tracker, and (4) vulnerability dashboard and food crisis risk monitoring panel. VAM1, VAM2, and VAM3 refer to, respectively, WFP's (1) VAM/HungerMap LIVE and nowcasting tool; (2) CARI; and (3) Market Monitor.

systems, which directly estimate degrees of food insecurity; and *agricultural market information early warning systems*, which focus on supply and market conditions that could endanger food security. In addition, we indicate how these systems relate to the monitoring of chronic food insecurity. Figure 3 (and Table A.1 at the end of the chapter) charts the most important early warning systems by type of information provided (food insecurity or agricultural markets) and frequency of monitoring (annual, weekly, or daily). We discuss these below.

ACUTE FOOD INSECURITY EARLY WARNING MECHANISMS

Existing EWEA systems, developed and run by several international agencies, humanitarian organizations, and governments, have been instrumental in monitoring acute food insecurity as well as the drivers of food insecurity spikes in local contexts. These systems have been useful in identifying the need for food assistance and other humanitarian and development aid to stave off the worst consequences of food crises. The Integrated Food Security Phase Classification/Cadre Harmonisé (IPC/CH), often referred to as the gold standard for classifying degrees of food insecurity,[7] uses a five-phase scale, with Phase 3 considered "crisis level," where emergency food assistance is needed; Phase 4 is designated "emergency," with urgent action needed to save lives; and Phase 5 identifies a "catastrophe," or famine (Box 2). The USAID-supported Famine Early Warning Systems Network (FEWS NET) incorporates IPC/CH indicators in its forward-looking analyses of populations at risk of acute food insecurity.

The warning systems that monitor current food insecurity conditions in countries identified as

BOX 2 THE INTEGRATED FOOD SECURITY PHASE CLASSIFICATION (IPC): CONSENSUS-BASED IDENTIFICATION OF FOOD CRISES

IPC sets a common standard and shared language for classifying the severity of acute food crisis situations using a five-phase scale, and provides information on the number of people affected and on the drivers of food insecurity.[1] IPC classifications at the country level are based on a convergence of evidence, which works from the premise that various unrelated sources and types of data can "converge" toward strong conclusions. The Cadre Harmonisé (CH) is the IPC-compatible measure applied to food security conditions in West Africa. We refer to IPC/CH as one entity in this chapter. The table describes the five phases of acute food insecurity and the type of priority action expected from governments and the international community for each situation.

IPC/CH acute food insecurity phase description and priority response objectives

Phase	Phase description and priority response objectives
PHASE 1 None/Minimal	Households are able to meet essential food and nonfood needs without engaging in atypical and unsustainable strategies to access food and income. ***Action*** *required to build resilience and for disaster risk reduction.*
PHASE 2 Stressed	Households have minimally adequate food consumption but are unable to afford some essential non-food expenditures without engaging in stress-coping strategies. ***Action*** *required for disaster risk reduction and to protect livelihoods.*
PHASE 3 Crisis	Households either have food consumption gaps that are reflected by high or above-usual acute malnutrition, or are marginally able to meet minimum food needs but only by depleting essential livelihood assets or through crisis-coping strategies. ***Urgent action*** *required to protect livelihoods and reduce food consumption gaps.*
PHASE 4 Emergency	Households either have large food consumption gaps that are reflected in very high acute malnutrition and excess mortality, or are able to mitigate large food consumption gaps but only by employing emergency livelihood strategies and asset liquidation. ***Urgent action*** *required to save lives and livelihoods.*
PHASE 5 Catastrophe/Famine	Households have an extreme lack of food and/or other basic needs even after full employment of coping strategies. Starvation, death, destitution, and extremely critical acute malnutrition levels are evident. (For Famine classification, area needs to have extreme critical levels of acute malnutrition and mortality). ***Urgent action*** *required to revert/prevent widespread death and total collapse of livelihoods.*

Source: Reproduced from FSIN and GNAFC, *2022 Global Report on Food Crises: Joint Analysis for Better Decisions* (Rome: 2022).

at-risk – including IPC/CH and FEWS NET – typically provide only annual or, at best, quarterly assessments, which international aid agencies consider too infrequent to adequately address acute situations. Both mechanisms rely on combinations of primary and secondary information sources to identify vulnerable populations according to the IPC/CH classification system. The Food Security Information Network (FSIN) and the Global Network Against Food Crises (GNAFC) are important users of these monitoring mechanisms. FSIN integrates data from IPC/CH, FEWS NET, WFP, and other sources to reach interagency consensus about acute food insecurity situations and reports its assessments in the semi-annual *Global Report on Food Crises*. The GNAFC, which brings together multiple donors and international and regional organizations, uses the report's findings to prioritize places for assistance.

While the various acute food insecurity early warning systems are similar, they take different approaches to generating alerts. IPC/CH looks at the *current* acute food insecurity situation and then projects improvement or deterioration based on evidence and consensus of expert opinion convened at the country level (including key stakeholders from national governments, the United Nations, and nongovernmental agencies).[8] FEWS NET projects how risk factors – such as rainfall, price changes, conflict, and harvest prospects – are likely to affect the extent and severity of acute food insecurity in the *near term*, typically an eight-month period.[9] FEWS NET's analysis further differs from IPC's in that it does not include a consensus-based process.

The Vulnerability Analysis and Mapping (VAM) instrument, which includes Food Security Assessments, a Market Monitor, and a Seasonal Explorer, is a central element of WFP's early warning mechanism. The VAM brings together assessments of household-level food security conditions, local and global market trends in food supply and prices, harvest prospects, and food security risks associated with geopolitical and economic shocks (VAM1 in Figure 3). The system now also experiments with mobile technology, artificial intelligence, and data analytics to facilitate near real-time food security monitoring across countries, accessible through the HungerMap LIVE.[10]

Like FEWS NET, WFP's monitoring mechanisms combine geospatial, economic, and household data to analyze food security in the organization's 80 countries of operation. For its acute food insecurity assessments, WFP uses its Consolidated Approach for Reporting Indicators of Food Security (CARI, referred to as VAM2 in Figure 3), which combines food consumption scores, economic capacity indicators, and data on livelihood coping strategies to generate snapshots ("nowcasts") of existing acute food insecurity.[11] Nowcasts are used to identify the need for emergency interventions, including where and how food assistance is needed. WFP's assessments are used as inputs to the acute food insecurity estimates of all the countries covered by IPC/CH, but also cover many more food insecurity contexts. As a result, WFP's global estimate for the number of people facing crisis-level or worse acute food insecurity is much higher, 349 million in 79 countries in 2022 compared with 205 million in 45 countries identified in the *Global Report*.

WFP's assessments feed into its internal corporate alert system to trigger early responses. At the same time, drawing on information from IPC/CH and VAM1, the Food and Agriculture Organization of the United Nations (FAO) and WFP also issue joint early warnings on acute food insecurity through quarterly identification of "hunger hotspots," with country-specific recommendations for anticipatory action and emergency response.[12] These early warnings are based on projections of populations at risk, considering the presence of natural hazards, conflict, displacements of people, and economic shocks that are likely to drive acute food insecurity to crisis levels or worse. In the projections for October 2022 to January 2023 (issued in September 2022), FAO and WFP identified 19 hunger hotspots with 195.5 million people projected to be at risk of seriously worsening acute food insecurity. Of greatest concern are Afghanistan, Ethiopia, Nigeria, Somalia, South Sudan, and Yemen, where some vulnerable populations are already in or projected to suffer famine-like conditions.

The hunger hotspot mechanism is a good example of existing interagency collaboration in early warning systems for acute food insecurity and could provide a way forward for avoiding overlap and duplication. Similar collaborative efforts are also needed to better align the monitoring of acute and chronic food insecurity – which is important for avoiding confusion about the two concepts and thus about the magnitude of food insecurity problems – and to support alignment of humanitarian and development action in the fight against hunger. Currently, mechanisms for monitoring chronic food insecurity are disconnected from the early warning mechanisms for acute food insecurity. Standard indicators of chronic food insecurity are the prevalence of undernourishment (PoU) and severe and moderate food insecurity as measured through the Food Insecurity Experience Scale (FIES).[13] As indicated in Figure 1, many more people face chronic shortage of adequate food intake than face acute food insecurity. At present, these estimates of chronic food insecurity are not suitable for early warning purposes as they are available only with a time lag of a year and only include national aggregates and averages (for this reason, they are not included in Figure 3). Moreover, conceptually, the undernourishment and FIES indicators differ starkly from those for acute food insecurity. These fundamental differences hamper any analysis of the extent to which the risk factors linked with acute food insecurity and those linked with chronic conditions coincide, and thus, also hinder the alignment of humanitarian and development efforts.

AGRICULTURAL MARKET EARLY WARNING SYSTEMS

Agricultural market early warning systems monitor market trends, such as global price and supply shocks, that affect the stability of food availability and affordability. In many countries, the agriculture sector is a key driver of food security, as it is both the main supplier of food and a critical source of income. Adverse shocks to food supplies – caused by weather calamities such as droughts, fluctuations in global and local food prices, policies such as export restrictions, or other problems – can have significant impacts on food security in vulnerable regions. Unlike most acute food insecurity early warning systems discussed above, several of these information systems use high-frequency or near real-time data to provide alerts of the risk of supply and price shocks (Figure 3).

Several of these mechanisms were developed in the wake of the 2007/08 food price crisis. The most notable is the Agricultural Market Information System (AMIS), an interagency platform launched in 2011 by the G20 Ministers of Agriculture.[14] AMIS provides monthly assessments of global food supplies and identifies needs for coordinated policy action (such as avoiding export bans or managing food reserves) in times of market uncertainty. The GEOGLAM Crop Monitor, a related mechanism, is an international effort to provide open and timely remote-sensing information on global crop-growing conditions and agroclimatic factors, both for major food import and export regions and for countries where food security is at risk.[15] Similar global information services are also provided by, for instance, CropWatch of the Chinese Academy of Sciences, which also feeds into GEOGLAM's early warning system.[16]

IFPRI's Food Security Portal is another important platform developed in response to the 2007/08 crisis. The Portal tracks food price volatility in international markets, and its Excessive Food Price Variability Early Warning System (FSP1 in Figure 3) provides alerts of above-normal volatility through a traffic-light system that is updated daily (see page on IFPRI monitoring tools following this chapter). This warning system has provided early signals of market tightness and of the impacts of export restrictions on global market prices (see Chapter 4). Several additional tools and dashboards have recently been added to the Portal, most notably a Food and Fertilizer Export Restrictions Tracker (FSP2), a Fertilizer Market Dashboard, a Domestic Food Price Monitor (FSP3), and a Vulnerability Analysis tool (FSP4). These were developed in response to the series of recent food system shocks caused by COVID-19, multiple climate-related disasters, and the Russia-Ukraine war.

Other agricultural market early warning mechanisms with long data series include the information on international and national agricultural markets and policies provided by the U.S. Department of Agriculture's Foreign Agricultural Service (FAS) and the International Grains Council's monitoring of grains, rice, and oilseeds market conditions, including daily publication of its Grains and Oilseeds Index. FAO provides regular updates on market conditions for internationally traded agricultural commodities, including through its monthly Food Price Index, and on domestic food market conditions and dependence on food imports through its Global Information and Early Warning System on Food and Agriculture (GIEWS). The GIEWS continuously monitors food supply, demand, prices, and other key indicators at global and national levels and provides basic information for the pre-identification of (potential) food crisis countries covered by IPC/CH and the *Global Report*.

GAPS AND SHORTCOMINGS

While both types of warning systems provide vital information, better integration among these existing systems, addressing key data and analytic gaps, and a redefinition of famine criteria would boost their usefulness for humanitarian and development responses.

NEED FOR INTEGRATION

Insufficient integration of existing agricultural early warning systems with the acute food insecurity warning systems is a critical shortcoming. Joint FAO/WFP Crop and Food Security Assessment

BOX 3 GLOBAL FOOD PRICE SPIKES AND FOOD CRISES

The global food price spike of 2007/08 caught major development donors by surprise. As prices spiraled upward, the uniform response from donors was to provide more food assistance. Donors also recognized that underinvestment in R&D in prior decades had contributed to slow productivity growth and price spikes, and they committed to increasing funding for agricultural research. Despite this commitment, global food prices spiked again during 2010/11. However, the focus on global price shocks left donors blind to the 2011 famine in Somalia – which was driven by conflict and repeated drought, and cost the lives of a quarter of a million people before adequate emergency aid arrived.

The international price surges of 2007/08 and 2010/11 have often been referred to as "global food crises." This jump to identify a food crisis reflects a common presumption that higher prices in world markets are directly transmitted to domestic markets, raising domestic prices and eroding food access for vulnerable households. However, such price transmission is mediated by many factors, meaning that a spike in global food prices may not be adequate grounds for identifying a food crisis.

Will this time be different? When the Russia-Ukraine war began, food and fertilizer prices spiked in international markets during February–May of 2022. Many observers saw this as a "global food crisis," and most of the response so far has focused on increasing humanitarian assistance. Yet, the impacts of COVID-19 and the war in Ukraine on global food markets have highlighted the diverse vulnerabilities of food supply chains and other factors contributing to inadequate resilience of vulnerable populations. Better monitoring and understanding of those risks and how they can drive up food insecurity and hunger should be a priority for improving early warning systems.

Missions, which analyze countries' agricultural production alongside household food security, and the link between the FAO's GIEWS and IPC/CH are among the few examples of connection points between these two types of early warning systems. Also, WFP's VAM/Hunger Map LIVE, IPC/CH, and FEWS NET draw on data from the agricultural market warning systems. At present, however, these links remain weak. AMIS, for instance, only monitors staple food price trends in global markets and does not provide alerts regarding how (potential) shocks may be transmitted to domestic food prices at the consumer level (such as those for flour and bread) to signal food security risks. This disconnect may underlie the mistaken tendency to immediately interpret global supply or price shocks as a "food crisis" (Box 3), without examining how local contexts may moderate their impacts.

GAPS IN DATA AND ANALYSIS

At least four critical gaps in the data and analysis used by early warning systems need to be addressed.

CONFLICT AND FOOD INSECURITY. Conflict is known to be a primary driver of acute food insecurity (and identified as such in the main EWEAs for acute food insecurity), and growing evidence on the two-way causal relationship between conflict and food insecurity is becoming available.[17] But major research gaps remain, particularly in fragile contexts and situations of extreme food insecurity (see Chapter 7). Conflict is inherently a complex and politically sensitive phenomenon, and studying food security in conflict situations is difficult. Yet, understanding how conflict and food security interact is vital for analyzing and forecasting future food security scenarios, as well as for conflict mediation, which is an essential part of comprehensive intervention strategies. Lack of clear insight into how climate change heightens the risk of conflict increases this challenge. CGIAR's Climate Security Observatory and Climate Security Dialogues dashboard could be instrumental in helping to fill this void and strengthening the analysis of drivers of acute food insecurity as undertaken, for instance, for FAO and WFP's hunger hotspot assessments.

LIVESTOCK AND FISHERIES PRODUCTION AND MARKETS. Existing early warning systems include almost no indicators to monitor conditions in livestock and fisheries production and markets. These are important to the livelihoods of many poor and food insecure people, as well as increasingly important components of diets.

TRANSMISSION OF GLOBAL AGRICULTUAL MARKET SHOCKS. Understanding remains limited of how, and to what degree, these shocks are transmitted to domestic food supplies and prices, and how this affects the food security of vulnerable populations (Box 3). The transmission of global shocks varies greatly across countries and commodities and, hence, responses cannot be enacted with a broad brush. For example, a recent IFPRI study shows that the supply shock to global wheat markets from the Russia-Ukraine war was only very partially and gradually reflected in domestic price surges in most countries, though the impact was dramatic in some (see Chapter 1, Box 1).[18] Such analyses of shock transmission from the global to the more local context should be part of agricultural market early warning systems and should facilitate linkages with food insecurity warning systems.

CONSISTENCY IN ESTIMATES OF ACUTE FOOD INSECURITY. As mentioned, the existence of different estimates of the global number of acutely food insecure people is a source of confusion, reducing the credibility of the numbers and making it harder to communicate the magnitude of the world's hunger problems – ultimately slowing responses. Country coverage is the main reason underlying the difference between the global estimates for crisis-level acute food insecurity of the *Global Report* (205 million in 2022) and WFP (349 million). Hence, while costly in terms of information gathering, expanding coverage of the consensus-based IPC/CH analyses to all of the nearly 80 commonly recognized food-crisis countries would lead to a convergence in country coverage and consistency in estimates of acute food insecurity.

AN ACTIONABLE DEFINITION OF FAMINE

A shared understanding of the situation on the ground is essential for an early warning system to trigger early action, most importantly where there is a risk of famine. However, the present protocol for declaring famine is no longer sufficiently operational in current contexts. The existing IPC protocol is designed to gather information in slow-onset emergencies, such as prolonged drought. However, in many of today's food emergencies, conflict is the major driver pushing people to the brink of starvation. In these conflict contexts, information to determine whether there is famine must be collected within just a few hours at most. A revision of famine criteria is needed to allow for a consensus-based judgment call without time-consuming collection of detailed survey data.

ACHIEVING REAL-TIME MONITORING OF RISKS AND VULNERABILITIES

While the current suite of early warning systems provides valuable food insecurity projections, addressing the shortcomings could improve predictions and make them more useful for early action. How do we get there?

First, a stronger, clearer analytical framework for weighing the various risks and assumptions that go into short-term forecasts of both agricultural market conditions and acute food insecurity would improve the quality of predictions. Assumptions about how current conditions may relate to risk of future food crises should be analyzed systematically. By focusing on risk factors, vulnerabilities, and resilience, such a framework would better inform responses that address both immediate needs (acute food insecurity) and structural conditions that determine vulnerabilities and coping capacity.

Second, for these early warning systems to support early action that builds resilience before a potential crisis occurs or a current food crisis intensifies, future food security outcomes must be estimated many months in advance to give decision-makers time to plan interventions (see Chapter 3). The need for early alerts, along with the complexity of the factors shaping food security outcomes, complicates the work of projecting food crisis risk to inform early action. Most current systems are not equipped for this, though FEWS NET and WFP's VAM have taken important steps toward forward-looking estimation of acute food insecurity based on risk analysis.[19]

Third, models should be developed for more integrated, quantitative EWEA systems. Some efforts in this direction are already being made. The World Bank's Famine Action Mechanism (FAM) was set up in partnership with IPC/CH and FEWS NET, as well as other organizations, to scale up anticipatory (preventative) and early emergency action for emerging food crises. The FAM has explored statistical analysis and machine learning to help predict crises and inform responses through a new financing modality (Crisis Response Window Early Response Financing, under IDA 19).[20] Although this initiative's approach to predicting crises in real time shows promise for the use of forecasting technology, it has not yet produced results that could credibly underpin early action. Nonetheless, the FAM has used the framework to inform country-level consultations in Afghanistan, Chad, Somalia, South Sudan, and Yemen that aimed to identify anticipatory and early action programming. These consultations led to Somalia's shock-responsive safety net,[21] but unfortunately, this response did not prevent a renewed famine warning from being issued for large parts of the country in 2022.

The FAM shares features with risk-contingent credit lines, like the World Bank's Catastrophe Deferred Drawdown Option,[22] and "forecast-based finance" (FbF)[23] schemes that are being implemented by some disaster relief and humanitarian assistance agencies, including WFP, to anticipate disasters and prepare for action.[24] These schemes support EWEA decision-making and trigger access to humanitarian funding for early action based on in-depth forecast information and risk analysis. Under an FbF plan, participating agencies agree in advance on the allocation of financial resources for early action, as well on the specific forecast threshold that will trigger release of those resources, and roles and responsibilities of everyone involved (see Chapter 3). FbF schemes are often deployed in well-defined fragile settings, for predefined beneficiaries and geographies. WFP's FbF scheme and other early action responses by governments and humanitarian aid agencies are informed by the HungerMap LIVE. Real-time monitoring through daily interviews with local informants conducted by call centers is active in 40 countries and complemented by predictive modeling that identifies food crisis risks for another 53 countries. The modeling uses acute food insecurity data as well as other relevant indicators, such as a rainfall-vegetation index, conflict reports, market prices, macroeconomic stability indicators, and nighttime light intensity (an indicator of economic development).[25] The near real-time nowcasts of the HungerMap build on machine-learning algorithms to yield information to monitor key drivers of food security risks and to make short-term forecasts of populations at risk of acute food insecurity. Compared with traditional information systems, real-time information makes it possible to identify deterioration in food security much more quickly, enhancing the early warning systems needed for anticipatory action and emergency response. Various research centers are engaging in similar efforts,[26] but these either focus only on a subset of risk factors or are as yet far from operational for use as part of EWEA systems. Much more work is needed to improve and tailor these efforts to support effective EWEA mechanisms that can inform concerted responses and align actions along the humanitarian-development nexus.

CONCLUSION

There is simply not enough funding available to address the increasing number of crises and beneficiaries already in need, and the deepening climate crisis will widen this funding gap. If interventions instead remain focused on emergency relief, the world will have to provide more and more assistance with every lean season. Better EWEA systems would allow governments and international agencies to tackle food crises earlier and more effectively, and to reorient interventions toward resilience building.

More work is needed to integrate – or at least to better explain – the different approaches and methodologies currently used by EWEAs to identify acute food insecurity, monitor chronic food insecurity, and incorporate the key information provided by agricultural market warning systems. More intensive cooperation between existing platforms, such as AMIS, on the one hand, and FSIN and GNAFC, on the other, will be needed to move forward. The importance of adequate agricultural market information and food security monitoring mechanisms has been repeatedly reiterated by multiple global platforms and gained prominence with the 2022 global food crisis, as reflected in the 2022 G20 Leaders' Declaration and the initiative of the G7 Development Ministers for a Global Alliance for Food Security.[27] This internationally concerted support is essential but will be most effective if it drives the improvement and integration of existing mechanisms rather than creating new ones.

To effectively and efficiently support responses to food crises, early warning systems must be enhanced to include real-time monitoring of key risk factors and vulnerabilities that affect food access in rapidly changing global and national contexts, along with analyses of how those risk factors and vulnerabilities increase the likelihood of food crisis situations. Most importantly, they must inform policy recommendations for buffering the most harmful impacts of those shocks in the short run and for building sustainable resilience for the long term. The chapters that follow explore both the types of early action that early warning systems can facilitate and short- and long-term policy responses that can reduce the frequency and impact of food crises.

TABLE A1 Description of early warning and other relevant monitoring systems

Name	Organization	Description	Link
IPC/CH – Integrated Phase Classification/ Cadre Harmonisé	Food and Agriculture Organization (FAO), World Food Programme (WFP), and multiple partners	The IPC provides decision-makers with core estimates of severity and magnitude of acute and chronic food insecurity and malnutrition using evidence and consensus-based analysis to inform emergency responses as well as medium- and long-term policy and programming.	http://www.ipcinfo.org/
FEWS NET – Famine Early Warning System Network	U.S. Agency for International Development (USAID)	FEWS NET provides unbiased, evidence-based analysis to governments and relief agencies that plan for and respond to humanitarian crises. Its analyses also support resilience and development programming. FEWS NET posts monthly reports on several dozen countries, primarily in sub-Saharan Africa.	https://fews.net/
VAM1 – Vulnerability Analyses and Mapping and Hunger Map LIVE	WFP	VAM1 provides vulnerability data for food security analysis and monitoring and real-time "nowcasting" of food security situations in more than 80 countries to inform WFP planning and resourcing.	https://dataviz.vam.wfp.org/ https://hungermap.wfp.org/
VAM2/CARI – Consolidated Approach for Reporting Indicators of Food Security	WFP	WFP's CARI provides "snapshots" of acute food insecurity situations based on multiple indicators, including food consumption scores, food energy shortfall, poverty status, food expenditure shares, and livelihood coping strategies. WFP uses this information to identify need for emergency interventions.	https://www.wfp.org/publications/consolidated-approach-reporting-indicators-food-security-cari-guidelines
VAM3 – WFP Global Market Monitor	WFP	WFP's monthly Global Market Monitor provides information on changes in the cost of basic food baskets, alerts for price spikes in local markets, and domestic inflation and currency movements as well as an overview of global food commodity price developments. Price information is publicly available and covers more than 1,500 markets.	https://www.wfp.org/content/market-monitor
GIEWS – Global Information and Early Warning System	FAO	GIEWS continuously monitors food supply, demand, prices, and other key indicators at global and national levels for assessing the overall food security situation in most countries of the world.	http://www.fao.org/giews/en/

Table A1 continued

Name	Organization	Description	Link
FAOSTAT	FAO	FAOSTAT provides free access to food and agriculture data for more than 245 countries and territories and covers all FAO regional groupings, from 1961 onward. It includes annual crop production, agricultural trade, and food balance sheets, among other data.	http://www.fao.org/faostat/en/#home
GEOGLAM – Crop Monitor of the Group on Earth Observations Agricultural Monitoring	GEOGLAM	GEOGLAM provides open, timely information on crop conditions in support of market transparency for the G20 Agricultural Market Information System (AMIS) as well as an early warning system for countries at risk of food production shortfalls.	https://cropmonitor.org/
CropWatch	Chinese Academy of Science	CropWatch assesses national and global crop production and related information using remote sensing and ground-based indicators.	http://www.cropwatch.com.cn/htm/en/index.shtml
AMIS – Agricultural Market Information System	AMIS (multi-agency, multi-country/G20)	AMIS is an interagency platform, composed of G20 members plus Spain and 7 additional major exporters and importers of agricultural commodities, to enhance food market transparency and policy responses for food market stability.	http://www.amis-outlook.org/
FAS – Foreign Agricultural Service	U.S. Department of Agriculture (USDA)	FAS links US agriculture to global agriculture and food supplies information to enhance export opportunities and provide information about prospects for global food security.	https://www.fas.usda.gov/
FSP1 – Food Security Portal (FSP) Excessive Food Price Variability Index	IFPRI	The FSP provides a real-time early warning system for price trends and price volatility in international markets for key agricultural commodities. International prices and the volatility index are updated daily.	https://www.foodsecurityportal.org/tools/excessive-food-price-variability-early-warning-system
FSP2 – Food Security Portal (FSP) Food and Fertilizer Trade Restrictions; Fertilizer Market; and Production and Stocks Trackers	IFPRI	The FSP provides daily updates of food and fertilizer trade restrictions, monthly updates of fertilizer prices and market conditions, and monthly updates of global supply and stocks of key staple foods.	https://www.foodsecurityportal.org/tools/COVID-19-food-trade-policy-tracker https://www.foodsecurityportal.org/node/1947 https://www.foodsecurityportal.org/node/1734
FSP3 – Food Security Portal (FSP) Domestic Food Price Monitor	IFPRI	The FSP provides a dashboard with trends in domestic food prices (aggregate and by main food items) with most prices updated monthly. It also includes a dashboard to track transmission of international price trends and other determinants of domestic food price inflation.	https://www.foodsecurityportal.org/node/2089
FSP4 – Food Security Portal (FSP) Vulnerability Analysis Dashboard	IFPRI	The FSP provides a dashboard identifying food insecurity hotspots and vulnerability to different types of global market shocks and other risk factors.	https://www.foodsecurityportal.org/tools/control-panel-for-risk-monitoring

TOOLS AND MODELS

FPRI has developed a wide set of tools to support early warning, analysis, and food crisis response. The Food Security Portal provides access to databases and alert systems that monitor changes in agrifood markets, fertilizer markets, agricultural production, and stocks as well as country-level vulnerability, and that track policy responses to crises. IFPRI also supports analytic models that provide projections of crisis and policy impacts.

EARLY WARNING AND DATA TRACKING TOOLS

FOOD SECURITY PORTAL
Facilitated by IFPRI and supported by the European Commission, the Portal provides food security alerts and open access to country-level data, analytic tools, and research products to help policymakers and other stakeholders respond to developments in the world food system. The tools described here, along with many others, are available through the Portal.
https://www.foodsecurityportal.org/

EXCESSIVE FOOD PRICE VARIABILITY EARLY WARNING SYSTEM
Identifies high volatility in markets, which can jeopardize food security, to alert farmers, traders, processors, and policymakers of upcoming uncertainty in prices for staple crops.
https://www.foodsecurityportal.org/tools/excessive-food-price-variability-early-warning-system

EARLY WARNING HUB
Brings together a number of major early warning systems from international organizations, providing information on the latest food security alerts and situations on the ground.
https://www.foodsecurityportal.org/tools/early-warning-hub

CONTROL PANEL FOR RISK MONITORING
Allows real-time monitoring of risk factors for food crises, including data on conflict, COVID-19, crop conditions, and climate-related satellite data.
https://www.foodsecurityportal.org/tools/control-panel-for-risk-monitoring

FOOD AND FERTILIZER EXPORT RESTRICTIONS TRACKER
Tracks records country policies that restrict food exports and chemical fertilizer exports that can contribute to food insecurity, and their likely impact in terms of US dollar value and kilocalories.
https://www.foodsecurityportal.org/tools/COVID-19-food-trade-policy-tracker

FERTILIZER MARKET DASHBOARD
Provides several datasets to monitor fertilizer markets including the latest monthly prices of major fertilizers and natural gas – a key input in fertilizer production – and country profiles of fertilizer trade and use.
https://www.foodsecurityportal.org/node/1947

PRICE SHOCK TOOL
Allows interactive exploration of the impact of commodity prices changes on poverty for a set of vulnerable countries.
https://www.foodsecurityportal.org/node/1627

PRODUCTION AND STOCKS MONITORING SYSTEM
Provides data on four major commodities, covering production, domestic consumption, ending stocks, and stock-to-use ratios by country and across time, to monitor vulnerability of world food markets.
https://www.foodsecurityportal.org/node/1734

MODELS

RIAPA
IFPRI's Rural Investment and Policy Analysis (RIAPA) modeling system offers forward-looking, economywide country-level analysis that serves as a simulation laboratory for different policies, investments, or economic shocks. RIAPA covers over 30 countries and includes detailed representation of agrifood systems that allows analysts to measure impacts to food systems, national economies, and global markets. A series of RIAPA analyses on the impact of the Russia-Ukraine war are found here:
https://www.ifpri.org/spotlight/food-prices-war-ukraine

IMPACT MODEL
To explore the long-term challenges policymakers face in sustainably reducing hunger and poverty, IFPRI developed the International Model for Policy Analysis of Agricultural Commodities and Trade (IMPACT) Model. IMPACT is a network of linked economic, water, and crop models, allowing it to account for environmental, biophysical, and socioeconomic trends in simulations of national and international agricultural markets. It covers 44 commodities that make up nearly all of the world's food production and consumption. Information on IMPACT and simulation results are found here:
https://www.ifpri.org/project/ifpri-impact-model

CHAPTER 3

Crisis Resilience
Humanitarian Response and Anticipatory Action

SIKANDRA KURDI AND SANDRA RUCKSTUHL

Sikandra Kurdi is a research fellow, Development Strategies and Governance Unit, International Food Policy Research Institute. **Sandra Ruckstuhl** is special advisor to the director general and a senior researcher, International Water Management Institute.

KEY MESSAGES

- The vast majority of humanitarian response is activated after a crisis occurs, delivering lifesaving aid, but at relatively high costs and in a framework that prioritizes short-term solutions over long-term resilience.
- Better evidence can help align humanitarian aid delivery with medium- and long-term development strategies and with resilience building.
- In the anticipatory action approach, costly delays and suffering can be avoided. Pre-allocating financial resources and preplanning responses to be activated when a trigger level is reached in a risk-monitoring system ensure efficient responses to crises.
- Beyond the narrow definition of anticipatory action as a preplanned emergency response, the broader conception of promoting resilience should guide policymakers in investing in long-term development goals even in fragile and conflict-affected contexts.

To improve the impact of humanitarian response and anticipatory action, it is important to:

- Increase data collection and analysis, including impact assessments, of humanitarian assistance and anticipatory action programs in different contexts, particularly in fragile and conflict-affected settings.
- Develop anticipatory action frameworks that pre-identify vulnerabilities and funding triggers, ensure regular data collection for risk monitoring, define clear roles and responsibilities, and identify available financial resources before crises hit.
- Assess the targeting of the humanitarian assistance to identify what groups are being missed and ensure their inclusion.
- Support interventions that reflect the humanitarian–development–peace nexus, such as nutrition-sensitive programming, use of local procurement, support for local institutions, and transitioning aid toward more permanent safety nets.

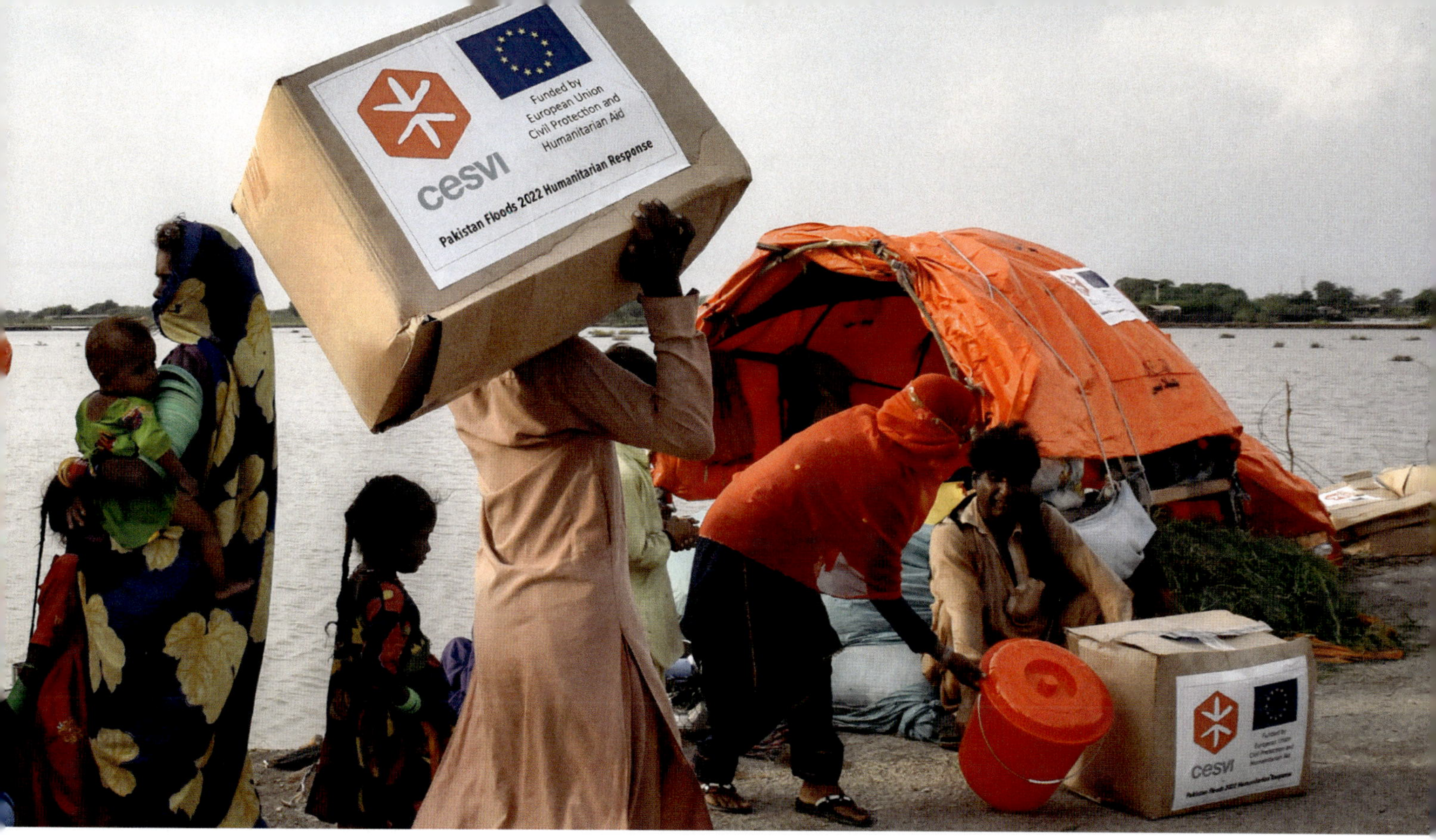

In human, economic, and environmental terms, the total cost of disaster and crisis response is extremely high, and the disastrous combination of the food price crises coming on the heels of the COVID-19 pandemic and natural calamities is straining public budgets and squeezing financial options. In 2020, private and public losses from weather-related disasters alone exceeded a total of US$258 billion globally – 29 percent above the 2001–2020 average – making it the fifth costliest year on record, and rising temperatures are expected to bring even more frequent and severe extreme weather events.[1] At the same time, conflict has become a leading contributor to humanitarian crisis situations – as seen most recently with the food and energy crises precipitated by the Russia-Ukraine war and refugee flows driven by the Syrian civil war.[2]

Timely response to crisis situations is critical. Households that have been displaced or lost their livelihoods can rapidly deplete savings and engage in coping strategies of last resort, which have long-term costs for well-being, with poor or near-poor households particularly vulnerable.[3] Even worse, shocks can stoke fragility, reduce effectiveness and inclusiveness, and displace standards of good governance, contributing to a perpetual cycle of instability. Institutions and researchers are increasingly grappling with finding the most efficient and effective ways to mitigate disaster costs through preemptive action, preparedness, and relief.

HUMANITARIAN AID FLOWS

Globally, the United Nations Office for the Coordination of Humanitarian Affairs (OCHA) reports that US$41 billion was needed to reach 183 million people targeted for international humanitarian assistance in 2022. Most people in need are living in countries affected by protracted crisis and conflict, with the largest numbers of targeted beneficiaries in Ethiopia (22.3 million), Afghanistan (22.1 million), Yemen (16.0 million), and Syria (12.0 million).[4] The gap between needs and funding has grown significantly since 2019, with only 46 percent of the global appeal funded in 2021, and international aid funding is not projected to keep pace with increasing need. Maximizing the efficiency of these aid flows is more important than ever.

Humanitarian response to crisis and disaster situations is grounded in principles of independence, neutrality, and impartiality, which grew out of longstanding concern about the risks of delivering aid in situations where the normal local political authorities are unable or unwilling to do so.[5] These foundational principles allow humanitarian actors to deliver lifesaving aid in extremely challenging circumstances, but by the same token, they constrain delivery mechanisms and operations in ways that prioritize meeting short-term emergency needs over building resilience and human development in the longer term.[6] For example, investing in local institutional capacity or procurement from local suppliers, both important for building resilience, invites questions about impartiality and independence, and programming that goes beyond the most immediate human needs for survival may generate controversy with local authorities about how longer-term goals are prioritized.

EVALUATING AID PROGRAM IMPACTS IN HUMANITARIAN RESPONSE CONTEXTS

Compared with social assistance programs in stable contexts, where research has long played a key role,[7] there is relatively little rigorous research on the impacts of assistance in humanitarian settings. Donors, practitioners, and the academic community have called for more rigorous evaluation of humanitarian assistance programs,[8] and researchers from the International Food Policy Research Institute (IFPRI) are major contributors to the small but growing body of evidence in humanitarian and crisis contexts.[9] Studying humanitarian programming specifically is important because lessons from stable contexts do not always carry over into settings where implementation is more challenging and where beneficiaries face more frequent and severe shocks. For example, the greater level of instability faced by beneficiaries in such contexts may substantially change household investment and risk preferences. Among a series of similarly implemented graduation programs, impacts on consumption were significantly lower in Yemen than in stable country contexts. This result may reflect difficulties with program implementation or conflict-affected households' greater desire to maintain assets (in this case, livestock) as a buffer stock for coping with future shocks.[10]

In a study on World Food Programme (WFP) emergency operations amid the conflict in Mali in 2013–2014, researchers showed that food assistance had a significant impact on micronutrient availability. The increased availability of food translated into gains for child height in areas less directly affected by the conflict, while in the villages most directly affected by conflict, the significant program impacts were on total household expenditures rather than on child nutritional status.[11] The study also showed that in areas of Mali most highly exposed to conflict, both general food distribution and school feeding programs led to increased school enrollment, but in areas less exposed to conflict, school feeding programs increased enrollment and educational attainment, while general food distribution was negatively associated with enrollment.[12] These results highlight how impacts of assistance can be affected by the specific emergency context.

While cash-based programs gained popularity in the developing world in the 2000s, cash-based programming for humanitarian responses has emerged as a growing trend only in the past decade. Cash transfers are easily scalable, fast to roll out, and usually considerably cheaper than in-kind assistance and less distorting of local production systems. IFPRI research, including several studies mentioned below, has been cited in good practice guidelines for the use of cash transfers in humanitarian response.[13]

As part of an ongoing partnership with WFP, IFPRI conducted a comparative analysis of cash, voucher, and food assistance using randomized controlled trials in humanitarian response contexts in Ecuador, Niger, Uganda, and Yemen.[14] Cash or vouchers were found to be more effective for improving dietary quality in most contexts, but food distribution generally had greater impact in terms of increasing calorie consumption. Yet the relative benefits of cash transfers or vouchers compared with equally valued food distribution varied considerably depending on the country, highlighting the need for research in a wide variety of contexts to provide relevant guidance to humanitarian operations.

Two other recent studies in Yemen highlighted the nutritional impacts of cash transfers supported by international aid: an emergency cash transfer program combined with child nutrition programming had significant impacts on child dietary quality as well as reduced stunting for the poorest households during the current crisis; and cash transfers during an earlier period of instability were associated with less wasting.[15]

ASSESSING THE TARGETING OF HUMANITARIAN AID

Another key challenge for humanitarian aid operations is how best to target relief efforts. Compared with development programs in stable contexts, humanitarian responders have far less administrative data, more mobile populations, and a much shorter timeline for identifying the neediest beneficiaries. Interagency evaluations of humanitarian relief operations in Ethiopia, South Sudan, and Yemen highlight challenges such as a lack of consolidated databases across agencies and NGO clusters, insufficient geographic targeting of aid due to difficulties with access, and perceptions (indicated by focus groups) that the selection of aid recipients is arbitrary or unfair.[16] While not all targeting is efficiently organized and trusted even in stable contexts, the greater local accountability for the implementing institutions in stable contexts may lead to more positive perceptions of the targeting process.[17] Rigorous assessment of targeting of humanitarian responses can clarify what groups risk being missed by existing methodologies. For example, an assessment of a food distribution effort in Ethiopia showed that, in contrast to the national social protection program, which targeted households in the poorer quintiles of the wealth distribution, local officials targeted humanitarian food assistance to households with more wealth, but which had experienced a negative shock in the past 12 months.[18]

"Shock-responsive" social protection programs solve many of the challenges of emergency targeting by leveraging existing programs and databases to increase assistance during crises (see Chapter 5). However, it is important to keep in mind that inclusion in national social protection programs may be biased against some of the most vulnerable, such as migrants, people lacking legal status, women, and ethnic minorities.[19]

LINKING HUMANITARIAN RESPONSE WITH LONG-TERM DEVELOPMENT

At the 2016 World Humanitarian Summit, the global humanitarian community recognized the importance of coordination and strategic thinking around the humanitarian-development-peace "triple nexus" of rapid response, long-term recovery and growth, and political stability.

In practice, the long-term development thinking that has been operationalized in humanitarian response includes: (1) ensuring that food relief is nutrition-sensitive to support long-term health; (2) prioritizing local procurement and processing of food used in relief operations; (3) strengthening local institutions such as schools and local NGOs as partners during aid delivery; and (4) designing emergency aid programs in such a way that they can develop into national safety nets.[20]

NUTRITION. In terms of nutrition-sensitive food aid, distribution of fortified foods targeted to young children as part of the relief response in emergencies has been shown to prevent major losses in nutritional status. Providing supplemental food items with key micronutrients to children under two years old and to pregnant and lactating mothers is particularly important to ensure nutritional adequacy for human development during the first 1,000 days of life.[21] Children who receive adequate nutrition will have better health and earnings in the future, contributing to long-run development well after the crisis that led to the food distribution has ended. IFPRI and WFP jointly developed WFP's nutrition-sensitive program guidance by designing and evaluating nutrition-sensitive programs across a range of sectors. This guidance was rolled out in 2017–2018 and is being followed up by continued collaboration to assess the impact of nutrition-sensitive programming.[22]

LOCAL PROCUREMENT. Another way to keep the long-run impacts in view when running emergency response operations is to prioritize local procurement when possible. Relying solely on imported staple foods for food distribution can risk distorting local agricultural markets by lowering the demand for locally grown food. This distortion not only harms local farmers, but in protracted crisis situations can also reduce farmers' incentives to invest in production of locally consumed food items.[23] An IFPRI evaluation of WFP's Purchase for Progress program – in which low-income farmers were contracted as suppliers and provided with storage facilities – found significant increases in revenue for the farmers in the program, achieved through both higher prices and greater quantities sold.[24]

LOCAL INSTITUTIONS. International disaster aid has the potential to either undermine or support local institutions. This is particularly concerning in weak states and conflict-affected contexts, where long-run recovery relies on the establishment of good governance. Examples cited by researchers of cases where aid undermined local governance include the humanitarian crisis in Haiti after the 2010 earthquake and the failures of state building in Afghanistan and Iraq.[25]

SOCIAL SAFETY NETS. In addition to creating shock-responsive safety nets pre-crisis, designing humanitarian aid to transition into a more permanent social safety net provides an opportunity for both strengthening local institutions and promoting longer-term development goals (see Chapter 5). For example, in Yemen, funneling emergency cash transfers through a preexisting social protection system has preserved national institutions and maintained a basis for eventual reestablishment of the system post-crisis.[26]

ANTICIPATORY ACTION

With the overriding focus on meeting immediate needs and maintaining access, humanitarian responders may not always be able to deliver aid in ways that minimize costs and maximize long-term development goals. But what if, instead of being organized on a tight timeline post-crisis, responses could be planned ahead of time? This is the goal of the anticipatory action framework.

Now being piloted in multiple contexts, anticipatory action aims to protect households and communities before disaster strikes. The approach seeks to use humanitarian resources more efficiently by pre-allocating them to be spent in ways that reduce the impact of anticipated disasters.[27] This means using early warning or forecasting tools combined with predetermined decision-making protocols to inform early action for timely emergency response at the local, national, and/or international levels (see Chapter 2). Triggers or thresholds are predefined within data and risk monitoring systems. Figure 1 illustrates how initiating actions to address a crisis after early warning signs are detected, but before the full weight of the shock is felt, reduces the peak humanitarian need compared to traditional post-crisis humanitarian response.

Without an anticipatory action framework, fundraising in emergency situations, while urgent, can be complicated. Public and private sector actors, responders, and donors will need to reconcile their own spending priorities in the context of humanitarian need and decision-making structures that may be inadequate. As a result, humanitarian operations may be slow to start or to reach necessary capacity,[28] and it can take weeks or months for humanitarian aid to reach people in need if the response is only started post-crisis, often worsening impacts. Potential bottlenecks include evidence and data challenges, organizational mandates and operational policies, risk tolerance, and security and access issues. The 2011–2012 Somali famine is a prime example. In this case, nearly 260,000 people died, more than half of whom were children under five years of age. Analysis shows that, despite clear warning signs, large-scale morbidity, mortality, and displacement were caused by delays in international aid.[29] This has sparked major debates and some changes in humanitarian aid policy and practice – including a critical view of early warning mechanisms that failed to generate a rapid response. Taking this into consideration, anticipatory action initiatives need to operationalize preplanned response protocols and resource distribution strategies so that needs are met before they become critical and so impacts are mitigated.

FIGURE 1 Benefits of anticipatory action framework

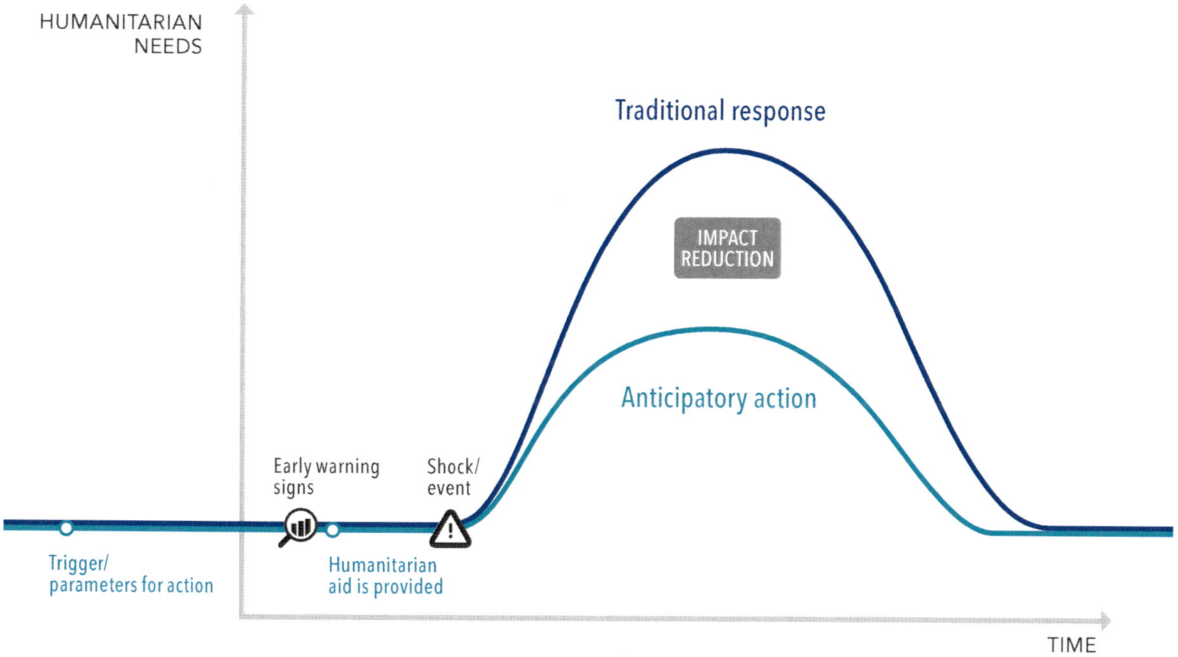

Source: Reproduced from OCHA Services, Center for Humdata, accessed February 1, 2023. https://centre.humdata.org/anticipatory-action/

For anticipatory action initiatives to be effective requires preparation in four areas:

PRE-IDENTIFIED VULNERABILITIES AND TRIGGER INDICATORS. Ensuring effective targeting and timely response requires an understanding of risks, exposure, and vulnerability in the particular context. With these clearly defined, monitoring systems can be more appropriately designed, using biophysical, social, and economic data to determine triggers for action. These types of mechanisms are especially challenged in very dynamic conflict- and migration-affected situations, where data on compound crises can be scarce and unreliable. In early applications, this approach was primarily used for weather hazards, but has now expanded to a wider range of risks such as epidemics and pests.

IMPACT-BASED RISK-MONITORING INFORMATION SERVICES. Risk monitoring requires regular data collection and calculation of updated risk levels using some of the approaches discussed in Chapter 2. Information services should be designed to ensure forecasts are impact-based, warnings reach the appropriate response agencies, and the vulnerable, and recipients understand how to respond.[30]

CLEAR ROLES AND RESPONSIBILITIES FOR DECISION-MAKING AND RESPONSE. Emergency responses may include, but are not limited to, cash subsidies and insurance, in-kind aid distribution, social protection services, humanitarian services and supply deliveries, and shelter. Roles, responsibilities, and procedures must be clear among all stakeholders involved in a humanitarian response initiative, and the initiative should be embedded within a broader disaster risk management and social protection strategy. This can be especially complex in fragile and conflict-affected settings, for example when government authority or capacity may be weak or nonexistent.

IDENTIFY AVAILABLE FINANCIAL RESOURCES AND RESOURCE MOBILIZATION STRATEGIES. The intent of anticipatory action is to establish data-informed decision-support systems to trigger quick disbursement of resources in emergency situations. Advanced planning can help identify needs and match financial resources with eligible beneficiaries, earmarking local and international resources and establishing disbursement processes.

INSTITUTIONAL INNOVATIONS

Despite broad agreement on the importance of planning ahead to mitigate crises, the structure of international humanitarian aid and government disaster response has not favored preemptive action. For example, some empirical evidence supports the idea that traditional post-disaster international aid creates a moral hazard problem – national governments that anticipate aid inflows are under-incentivized to invest in disaster mitigation.[31] Lack of coordination between agencies or ministries at both the national and international levels and between those responsible for emergency response and long-term investments has also been blamed for the lack of attention to anticipatory action.[32]

Recent institutional innovations, however, are poised to facilitate funding for anticipatory action at the global level. The UN Food Security Cluster Anticipatory Action Task Force has called for more donor funding to be dedicated to flexible uses or anticipatory actions,[33] and the UN's Food and Agriculture Organization has initiated several anticipatory action pilot projects with a total budget of US$6.2 million in 26 countries.[34] In May 2021, the G7 Foreign Ministers announced a commitment to "making the humanitarian system as anticipatory as possible" through both existing pooled funds and new financing solutions. Small-scale anticipatory components have already been added to the UN Central Emergency Response Fund, the International Federation of Red Cross and Red Crescent Societies' (IFRC) Disaster Relief Emergency Fund, and the World Bank's International Development Agency (IDA) Crisis Response Window.

Anticipatory action mechanisms are based on an action plan that is approved in advance and includes an agreed trigger for releasing funding to enact the plan, related to the expectation that a crisis is imminent. For example, the Forecast-based Action component of the IFRC Disaster Relief Emergency Fund launched in 2018 provides ready-to-go financing that can be released by early action protocols when triggered by forecasted natural disasters, such as hurricanes, floods, cold waves, and volcanic eruptions.[35] The IDA Crisis Response Window similarly provides funding conditional on reaching a trigger point for enacting a previously prepared Food Security Crisis Preparedness Response Plan.

HOW MUCH AND WHAT TYPES OF ANTICIPATORY ACTION ARE COST-EFFECTIVE?

The design and operation of anticipatory action initiatives are highly contextual. Research on the efficiency, effectiveness, and impact of these schemes is scant, and there are especially few examples of initiatives incorporating conflict prevention, mitigation, and peacebuilding. Inherent challenges arise in evaluating anticipatory action schemes and, because of the relative novelty of this approach, indicators and evidence of success are still being defined. Data collection is challenging in quick-onset disaster situations, and the time period over which the relative costs and benefits are expected to be calculated can be extremely long. But some attempts have already been made to collect experiences with anticipatory action and evidence to evaluate this approach.[36]

One clear benefit is the time savings in deploying humanitarian response operations. Action plans that include pre-positioning relief supplies, training first responders, and developing contingency plans for specific expected disasters can potentially allow relief to reach intended beneficiaries with better targeting, at greater speed and lower cost, and in ways that are better integrated with local markets and institutions. Significant savings in both time and cost have been found in practice when the IFRC used anticipatory action approaches to flooding in West and Central Africa and when WFP pre-positioned essential commodities for distribution in several countries.[37] The cost-savings free up resources for long-term adaptation investments, providing an incentive for donors to advocate for the broader establishment of anticipatory action systems.

Another way to measure the benefits of anticipatory action is to look at the degree to which earlier responses serve to protect long-term household and social welfare by reducing reliance on negative risk-coping mechanisms in the short term. For example, the short-term impact of drought on households may be income loss for farmers and production losses from crops and livestock, while long-term impacts include negative health effects, greater gender disparities, and reduced education, as well as increased migration, conflict, and political instability. Attempts to quantify such costs even at the level of aggregated estimates using an approach such as BACI (benefits of action–cost of inaction) can be informative about the potential for long-term savings from early investment in anticipatory action.[38]

Some limited quasi-experimental evidence on forecast-based financing provides more concrete measures of the gains from anticipatory action. Forecast-based financing is a type of anticipatory action in which distribution of aid is conditional on the forecast of an imminent crisis (see Chapter 2). A study of an IFRC forecast-based financing program in Mongolia showed that herder households who received assistance prior to an extreme winter season lost less livestock than households that did not receive assistance.[39] A qualitative study of a similar program in Bangladesh, which delivered government aid to communities identified as most likely to experience flooding in the upcoming season, found that beneficiary households used the cash to maintain food consumption and fund evacuation costs.[40]

PROMOTING CRISIS RESILIENCE

Anticipatory action shifts humanitarian funding availability from the response phase to an earlier point in time when it can be used for resilience building. In some models of anticipatory action programs, emergency funds reach individual households before a crisis hits, allowing those households to make investments that protect their livelihoods and assets. In other cases, the emergency funds are not distributed directly to households, but are used by local governments or other humanitarian actors to make investments in time to mitigate the worst effects of the crisis.

While anticipatory action is usually narrowly defined as a financing mechanism that is released based on a predefined forecast condition, investing in resilience to crises is also a broader concept. Resilience is most commonly understood as the ability to withstand and recover from external shocks, ensuring that short-run shocks do not have long-lasting adverse consequences. A wide variety of development goals – such as decreasing poverty, increasing access to basic services and education, improving institutions, and, at the household level, investing in productive assets and physical and mental health – can be viewed not only as ends in themselves, but also as means to improving households' capacity to absorb or adapt to shocks, as demonstrated by a large and growing body of research.[41] Despite this ongoing work, important knowledge gaps remain and new questions have emerged from the most recent crises.

GENERATING RIGOROUS EVIDENCE

Monitoring, evaluation, and impact assessment (MEIA) remains a major gap in anticipatory action and, more generally, in humanitarian and development interventions in fragile and conflict-affected settings (see Chapter 7). More evidence is needed on the impacts of different types of humanitarian assistance in different contexts, particularly in the most challenging places, and on targeting and shock-responsive social programming, integrating emergency aid with long-term resilience, and developing effective anticipatory action programs. More research is also needed on how to measure the cost-benefit ratio of investing in resilience. Operationalizing anticipatory action approaches requires work on building data sources to measure risks and on organizing stakeholder coalitions. A library of good practices, in addition to guidance for feasible and relevant MEIA techniques for anticipatory action, is needed to help develop and inform crisis responses. To this end, a new CGIAR Research Initiative on Fragility, Conflict, and Migration will implement a work program aiming to strengthen anticipatory action in complex crises and provide guidance to humanitarian programming on building long-run resilience. With evidence from this research program, policies can be implemented to reduce human suffering in the wake of natural disasters and conflict events.

CHAPTER 4

Agrifood Value Chains
Building Resilient Food Systems

BART MINTEN, BEN BELTON, AND THOMAS REARDON

Bart Minten is a senior research fellow and leader of the Myanmar Strategy Support Program, IFPRI. **Ben Belton** is a research fellow, South Asia Regional Office, IFPRI, and an associate professor of international development, Michigan State University. **Thomas Reardon** is university distinguished professor of agricultural, food, and resource economics, Michigan State University, and a nonresident research fellow, IFPRI.

KEY MESSAGES

- Agrifood value chains have transformed in recent years through rapid growth, increased diversity and complexity, and a revolution in logistics, storage, and retail. But they remain vulnerable to a variety of crises, including disease, conflict, and natural disasters. Their resilience varies with the type of shock, the structure of the chain, and the local context.
- Value chain impacts can evolve over the course of a crisis. During the COVID-19 pandemic, agrifood value chain actors first dealt with lockdowns, then with a downturn in demand, and finally with rising prices.
- Civil strife, conflict, and natural disasters disrupt food production and markets, often leading to rising food prices. Risks to food security and livelihoods can be reduced through flexible market mechanisms to support value chains as well as appropriate farming techniques and new insurance tools.
- Small, informal enterprises and women-owned enterprises are often more vulnerable to crisis impacts, as are producers and enterprises with limited market options.
- Agrifood actors respond to crises with short-term coping strategies and long-term adaptations. Improving coping strategies and pursuing transformation that facilitates adaptation are central to building resilience.

To ensure agrifood value chains contribute to recovery and resilience, it is crucial to:

- Tailor crisis response to the type of shock, the particular context and value chain, and when possible, different enterprise sizes.
- Invest in improved and innovative technologies and tools that build resilience, such as climate-smart agriculture and index-based insurance.
- Create a regulatory and business environment that fosters the development and widespread adoption of value chain innovations, such as e-commerce.
- Provide opportunities to continue private trading during crises, for example by avoiding trade restrictions and creating safe corridors.
- Ensure that women are able to take advantage of financial and digital innovations and have viable coping strategies.
- Conduct careful and frequent monitoring before and during crises to target assistance to crucial value chain nodes.

Agrifood value chains in the world's low- and middle-income countries (LMICs) have expanded rapidly over the past decade, supplying an increasing volume and diversity of food products. This transformation has been driven by the fast growth of urban and peri-urban areas as well as increasing demand from richer and more urban consumers for different, higher-quality, and often more expensive food. More farmers than ever are now connected to agrifood value chains through markets for both agricultural inputs and outputs. These connections are not only increasingly numerous but also increasingly complex, reflecting the greater diversity of products, inputs, and services that farmers buy and sell. Midstream and downstream in agrifood value chains, a "quiet" revolution has occurred in logistics, storage, transport, wholesale, retail, and food services, with fundamental and rapid changes in the structure, conduct, and performance of these value chain segments and the enterprises involved.[1]

This transformation provides new opportunities for farmers to increase their income and food security and nutrition,[2] and is generating revenues for technology upgrades, improving access to productivity-enhancing inputs and services, and increasing off-farm employment in small and medium enterprises (SMEs) in all value chain segments. At the consumer end, the expansion of value chains is essential for feeding urban residents, who now account for more than half the population in LMICs. But this transformation also brings new challenges. As value chains have become longer, stretching from rural areas to cities and across countries, they have also become more vulnerable to shocks that disrupt markets, including crises created by climate change, disease, and conflict. All these shocks can have major impacts for actors throughout the value chain, and consequently for livelihoods and consumers' food security.

Some research has begun to look at understanding, anticipating, and alleviating the impacts of crises on agrifood value chains. For example, researchers have identified five potential "hotspots" – aspects of value chains that can be particularly vulnerable to crises: (1) physical infrastructure (such as transport and storage), which affects risks to production; (2) geographic length of the supply chain, which affects potential disruptions such as road washouts along a supply route; (3) perishability of the product, which can make it vulnerable to delivery delays; (4) prevalence and robustness of physical capital, such as storage bins and cold chambers used by traders; and (5) "stranded assets," that is, assets that are only profitable in a particular end-market that the owner cannot access because of a supply chain shock.[3] Beyond these particular vulnerabilities, supply disruptions are determined by the nature and intensity of the shock and the exposure of supply chain actors.[4]

Minimizing the inevitable disruptions will require appropriate policy environments and investments all along the value chain, from the farm to the consumer. In the face of increasingly frequent shocks, the ongoing transformation will be essential to improving the adaptive capacity of agrifood value chains. To boost resilience, governments will need to create a business environment that fosters adaptation and innovation. In the private sector, continued investments in assets and good practices both in input supply chains (such as agro-dealers who provide inputs to farmers) and in the midstream of value chains (including processors, logistics firms, and wholesalers) will be essential to supporting food security during crises.

In this chapter, we review some of the recent evidence on the impact of three different types of crises – pandemics, conflict, and climate change – on the functioning of agrifood value chains and distill some lessons learned for building resilience.

THE COVID-19 PANDEMIC

COVID-19 – and the policies implemented to contain it – constituted an unprecedented shock to value chains worldwide. The challenges faced by agrifood businesses evolved over the course of the pandemic, reflecting policy shifts, the evolution of the disease, and changing economic conditions. In LMICs, threats to value chains progressed from mobility restrictions during the initial lockdowns, to depressed demand as economic activity declined, and most recently to price inflation.

SHOCKS TO VALUE CHAINS

During the first half of 2020, restrictions on transport and human mobility plus temporary closures of businesses and public institutions disrupted the flow of goods and services along global, regional, national, and subnational supply chains. For example, lockdowns prevented or delayed delivery of produce from farms to markets and inputs from factories to farms. In many cases, these restrictions raised food prices, at least temporarily. The initial containment policies implemented in developing countries often failed to consider the critical role of domestic supply chains in national food systems; in sub-Saharan Africa and India, for example, these supply chains deliver approximately 80 percent of the food consumed (by value).[5] As a result, lockdowns caused major disruptions to the food supply in many African and Asian countries and for diverse food commodities.

The impacts of initially stringent lockdown policies on transport and mobility were mainly short-lived, however. Businesses soon introduced work-arounds, such as operating on new routes or opening outside of normal business hours, and governments made quick policy adjustments, often prioritizing movement of agrifood products and farm inputs along with medicine and other essential products. Yet despite these adaptations, movement restrictions and related constraints on accessing materials and labor meant that many consumers experienced declines in income and loss of employment, which led to a drop in demand for food products. These impacts affected agrifood businesses directly. For instance, in Nigeria, a survey of enterprises in poultry and fish supply chains found that the main problems early in the pandemic were access to inputs, transport, and markets, along with low consumer demand.[6]

Over time, this set of challenges faded, while rising input costs and financial constraints (that is, inability to access or recoup loans) became more troublesome.[7] In many places, the combination of reduced consumer demand with rising input and operating costs squeezed the profits of farms and other supporting businesses, causing them to reduce their output or turnover.[8] Food prices rose as a result of these pandemic impacts, reaching the highest levels in a decade by the end of 2021, before rising even further in 2022 when the Russia-Ukraine war put additional pressure on global fuel and food prices.

RESILIENCE AND VULNERABILITY

Some enterprises were more vulnerable to these COVID-19 disruptions than others, and impacts were often context-specific. Women-led agrifood enterprises were sometimes, but not always, less resilient in the face of pandemic lockdowns than enterprises led by men, varying by country and by product. In Nigeria, for example, women-owned enterprises in poultry and fish value chains were 11 percent more likely to close than those owned by men between 2020 and 2021.[9] However, a similar study of SMEs in the midstream of potato and fish value chains in Kenya found no significant effect of the owners' gender on changes in business turnover during the pandemic.[10] And a study of agrifood SMEs in 17 countries found that although women-owned firms were more likely to report a production decrease of 30 percent or more as a result of the lockdowns, no differences were reported in business earnings between firms owned by women and men.[11]

Small, informal firms were sometimes, but not always, less resilient than larger, formal-sector enterprises. As with gender, the findings for firm size and COVID-19 impacts varied by context. In some countries, food enterprises in the informal sector (many of which are small and many owned by women) were disproportionately harmed by containment policies that favored formal businesses such as supermarkets.[12] In Senegal, large, formal, export-oriented vegetable farms fared better under COVID-19 restrictions than small farms and traders supplying domestic markets,[13] while in Nigeria, larger businesses in poultry and fish supply chains were 13 percent less likely to close than small businesses. On the other hand, smaller vegetable farms in Ethiopia were found to be *less* vulnerable to COVID-19 disruptions than medium-sized farms, because the smaller farms were less reliant on hired labor.[14] This pattern was also identified in Kenya, where smaller firms in the midstream segments of potato and fish value chains proved more resilient (as measured by relative changes in volumes traded) to shocks over the 2019–2021 period than larger businesses, though the very largest businesses surveyed experienced smaller relative reductions in sales.[15]

During the early stages of the pandemic, shorter supply chains (in terms of distance from farms to consumers) were expected to be more resilient than longer ones,[16] but their response has proven more complex. In Ethiopia, vegetable farmers who faced less competition from other areas (whether domestic or international) due to pandemic-related trade restrictions benefited through higher prices for their produce, while those who could no longer access markets within the country fared worse.[17] Among Australian agrifood businesses, those with both domestic and global value chain partners proved more resilient than those with only global business partners.[18] However, even in highly export-oriented sectors, firms with multiple potential buyers tended to weather the crisis well. For example, the Norwegian salmon industry was able to redirect products to alternative national markets and target alternative market segments by changing product forms, such as from fresh to frozen fish.[19] In contrast, supply chains delivering products to a single end-market, whether domestic or for export, were likely to suffer serious disruption, particularly where gluts of seasonal production coincided with movement restrictions, as happened with potatoes in Kenya.[20]

RESPONDING TO THE CRISIS

To address the diverse challenges related to the pandemic, businesses across all segments of agrifood value chains made two broad sets of adaptations: (1) short-term coping strategies, such as pausing or reducing business activities, minimizing operating costs, drawing down savings, or borrowing; and (2) longer-term or more proactive adaptations, such as the adoption of digital technologies, operational diversification, or increasing use of contracts (for example, retailers and processors signed contracts with farmers to reduce market risk for both parties).

Deep and abrupt changes in business practices, products, or technologies that substantially alter supply chains have been termed "pivoting."[21] Pivoting may be pursued as a short-term coping mechanism or as a long-term adaptation action and strategy. Pivots by food industry firms during the COVID-19 pandemic were focused primarily on marketing channels (such as shifting from in-store or in-restaurant sales to e-commerce) and technologies (such as shifting from labor- to capital-intensive

FIGURE 1 E-commerce yearly growth rates in 2019 (before COVID-19) and 2020

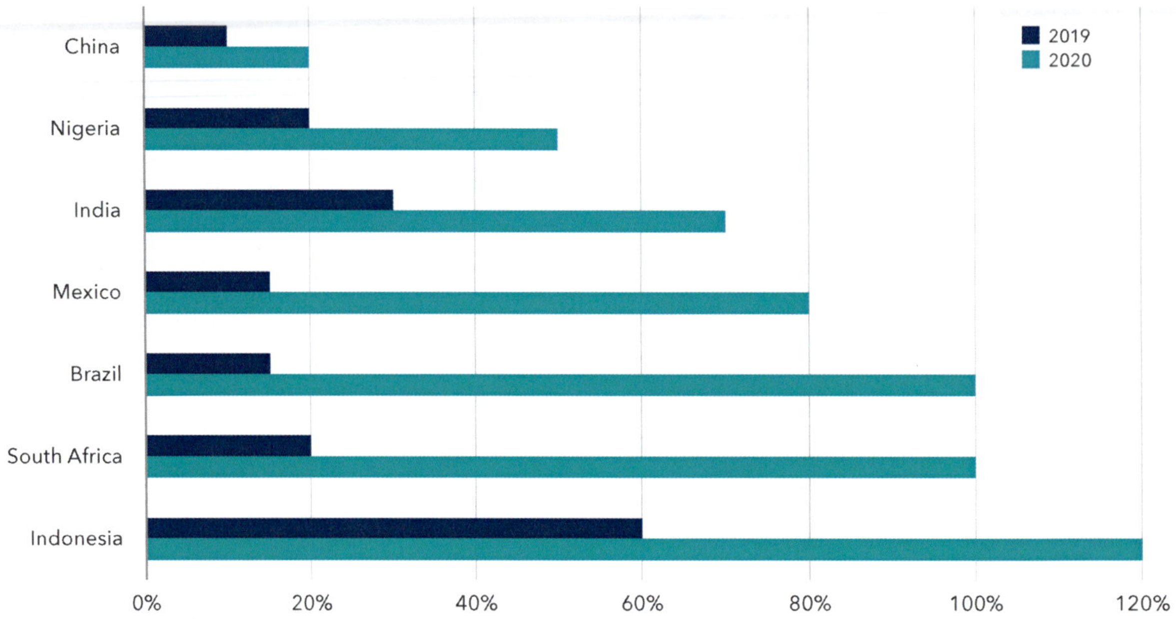

Source: Data from V. Vardhan, "Impact of the COVID-19 Pandemic on Retailing in Emerging Countries," Powerpoint presentation by Euromonitor International, October 2020, cited in T. Reardon, A. Heiman, L. Lu, C.S.R. Nuthalapati, R. Vos, and D. Zilberman, ""Pivoting' by Food Industry Firms to Cope with COVID-19 in Developing Regions: E-commerce and 'Copivoting' Delivery Intermediaries," *Agricultural Economics* 52, 3 (2021): 459–475.

technologies). Perhaps the most significant of these pivots was the boom in e-commerce for food retail sales in many LMICs, as consumers sought to prevent infection by avoiding public places. Although food e-commerce and associated online platforms and logistics infrastructure were growing prior to the pandemic, most notably in China, the pandemic dramatically boosted their expansion across multiple regions, albeit unevenly, with rapid growth in parts of Latin America, Southeast Asia, and India, and slower growth in much of sub-Saharan Africa (Figure 1). The differences in regional expansion are explained by the basic enabling conditions for these businesses in the different regions – hard infrastructure like roads that allowed fulfillment of digital orders, and soft infrastructure like business regulations that did not fetter the establishment of new enterprises in this sector.

Uptake of digital information and communication technologies (ICT) by smaller actors and individuals in food value chains also accelerated significantly. For instance, in Kenya, the use of "mobile money" for making transactions and of phones, social media, and online marketplaces to search for buyers or sellers increased in the wake of the pandemic.[22] Similar changes, though starting from a lower base, have also been observed in Nigeria.[23] In India, accepting digital payment via QR codes became very common, even among the smallest retail businesses.[24]

Signs of a partial business recovery were evident in most countries in 2021 despite the severity of the Delta variant of COVID-19, which predominated by mid-year. In LMICs, this business resilience reflects in part the nature of the small farms and firms that make up the bulk of the food system. These enterprises typically have low overhead and few hired workers, which allowed them to scale back operations and persist through times of crisis. However, their coping strategies, including drawing down savings and borrowing, likely exacted a heavy cost in terms of human welfare and eroded their capacity to adapt to future shocks. The largest businesses have been best placed to adapt proactively through pivots – such as the reconfiguration of supply chains and diffusion of e-commerce – often facilitated by co-pivots by other supply chain actors such as logistics providers.[25] These disparities may have contributed to increasing concentration in ownership and market shares in some instances.[26] However, many smaller farms and firms were also quick to adopt or increase their use of ICT to overcome coordination problems and reduce the need for physical contact during the crisis, and they look set to continue on this path.

CIVIL STRIFE AND CONFLICTS

Most of the world's extreme poor live in fragile states.[27] Yet despite the enormous importance of these areas for global food security, relatively little research has examined how agrifood value chains respond and adapt in such contexts. Food prices and the affordability of food are particularly critical in these settings. Some research has shown that rising food prices are often the cause of violent conflict and unrest, and there are significant linkages and feedback loops between violence and food prices. Here we briefly discuss two conflicts – one international (the Russia-Ukraine war), which has global implications, and one with largely domestic impacts (Myanmar).

RUSSIA-UKRAINE WAR

The recent war in Ukraine has highlighted the vulnerability of global agrifood value chains to conflict. International commodity markets were already volatile before the war began, as a result of the COVID-19 pandemic and weather shocks that reduced harvests. Food prices were at their highest level in a decade, and international fertilizer prices had increased by 125 percent in the preceding year, due to high prices for natural gas and coal (used as feedstock and energy sources in ammonia production).[28] High energy prices and a global shortage of containers also led to a substantial increase in international shipping costs in 2021. When Russia invaded Ukraine in February 2022, food prices increased rapidly, especially wheat prices, which rose by more than 40 percent in just two months. By July, they had returned to pre-conflict levels, but to date remain well above the historical average. Fertilizer prices also increased, given that Russia and Belarus were major fertilizer suppliers,[29] which has sparked serious concerns about food security, especially in LMICs.

While the countries most dependent on wheat imports from Ukraine and Russia were directly affected, the impact of the war on agrifood value chains has been global, affecting many LMICs that import wheat and leading to spillover effects on other value chains.[30] As with the food crises in 2007/08 and 2010, some countries have tried to shield themselves from rising food prices by implementing export restrictions or lowering import restrictions. These trade policy interventions only aggravate the global problem by escalating disruptions of global agrifood value chains and food price volatility.[31] Price-insulating policies adopted by a number of countries in the wake of the Ukraine crisis have contributed to high volatility in world prices, as price risks were transferred from one group of countries to another.[32] However, some unambiguously beneficial trade policies have also been adopted in the crisis, such as the creation of safe corridors that can help reduce the impact of the conflict – including the Black Sea Grain Initiative, which has allowed grain exports from Ukraine's seaports.

CONFLICT IN MYANMAR

In Myanmar, civil strife has disrupted domestic agrifood value chains. Myanmar's military seized control in a coup in February 2021, setting the country on a path toward widespread violence, insecurity, and major economic contraction. In protest, Myanmar's people organized a national Civil Disobedience Movement and worker strikes that disrupted service delivery for both public institutions and private businesses. Banks discontinued in-person services and faced severe liquidity shortages, limiting businesses' ability to pay employees and suppliers as well as individuals' access to their money. Cumulatively, these disruptions had major economic consequences – GDP declined by 18 percent and the poverty rate increased by between 8 and 18 percentage points.[33]

The value chain for rice, Myanmar's primary staple food, is the country's biggest and is closely linked to the banking and transport sectors. In the aftermath of the coup, a number of challenges arose in the rice value chain, as banks were short of cash and transport was complicated by lack of drivers, lack of fuel or high costs of fuel, and road blocks.

Nonetheless, rice processing and trade continued, ensuring that rice was available in most retail markets, and processing margins remained largely stable – demonstrating the value chain's resilience to such major shocks.[34] This resilience reflects the ability of value chain actors to adapt to new conditions. For example, to address the impact of the banking crisis, millers began using a modified *hundi* payment system to sell rice, whereby a sale is negotiated and payment is transferred from a trusted third party with available cash. However, as the margin increased between the price that rice millers received and the price that retailers charged consumers, average retail prices rose by 11 percent, implying welfare losses of almost US$500 million for the country (equivalent to 3 percent of agricultural GDP). Despite the knot of problems that must be addressed in such settings, there are policy tools that can help. For example, easing transport restrictions and facilitating cheap and safe trade of food products can reduce food price inflation, ensure higher farm prices, and thus improve welfare.[35]

NATURAL CALAMITIES

Natural calamities, such as floods and droughts, cause major disruptions in agrifood value chains, as the recent disastrous flooding in Pakistan has reminded the world. Such extreme weather events are occurring with greater frequency, and the shocks can affect a wide area. For example, floods in Bangladesh in 1998 covered two-thirds of the country, causing severe damage to the country's rice crop. However, widespread food insecurity was avoided, as Bangladesh's rice markets adjusted to the loss of domestic production through significant commercial rice imports from India. Because Bangladesh had liberalized trade in the early 1990s, private traders were assured that sufficient rice imports would be available.[36] In Ethiopia, where drought is common, the worst impacts may be avoided by adopting appropriate agricultural practices. A study found that training farmers in the production and conservation of livestock fodder as well as in soil and water conservation practices – good practices even in normal times – was crucial for strengthening farmers' capacity to adapt to and cope with drought.[37]

The increasing frequency of natural calamities that affect agrifood systems has generated significant interest and experimentation with innovative index-based agricultural insurance products to reduce the risk faced by farmers. While globally about half of all farms are covered by agricultural insurance, a substantial number of farms in LMICs are left out.[38] Index-based insurance products are generally perceived to be too expensive for smaller farms in these settings, and uptake has been low. Given the importance of such risk-reducing products in increasing the resilience of agrifood value chains, one promising option for improving uptake is to bundle these insurance products with stress-tolerant seed varieties, risk-oriented credit/savings products, or extension services.[39] However, despite innovations and new opportunities created by advances in remotely-sensed data systems, digital technologies, smartphones, and e-banking, insurance for catastrophic risks is expected to remain unaffordable for most farmers and thus is an inequitable form of safety net,[40] leaving many farmers dependent on disaster assistance in catastrophic years.

LESSONS LEARNED

A number of lessons can be drawn from the responses of agrifood value chains to these different crises.

AGRIFOOD VALUE CHAINS HAVE GENERALLY PROVEN QUITE RESILIENT TO SHOCKS, THOUGH IN WAYS THAT ARE HETEROGENOUS AND CONTEXT SPECIFIC. The type of crisis – driven by climate, civil strife, pandemic, or other shocks – affects value chains' resilience. The resilience of agrifood value chains has sometimes come from direct policy interventions, as seen in exemptions of food service industries from lockdowns in the case of the COVID-19 pandemic. Structural differences in value chains can make one value chain more resilient than others. In particular, small informal firms may be more vulnerable or face greater constraints in their response than large formal firms. Thus, interventions and policies should be tailored not only to the type of crisis but also to the specific context, value chain, and if possible, size of the enterprise affected.

IMPROVED AND NEW TECHNOLOGIES HAVE AN IMPORTANT ROLE TO PLAY IN ENSURING GREATER RESILIENCE OF VALUE CHAINS. Proactive investments are needed to establish widespread availability of usable knowledge and shock-resistant technologies, such as climate-smart technologies and practices, and relevant ICT. In the past, breeding efforts by international research organizations have led to lower yield volatility,[41] and today the development of new agricultural technologies again has an important role to play in improving risk management. In addition, appropriate insurance instruments and risk mitigation strategies should be facilitated, with public sector interventions playing a crucial role.

A REGULATORY AND BUSINESS ENVIRONMENT NEEDS TO BE CREATED (OR EXISTING ENVIRONMENTS REFORMED) TO ALLOW SHOCK-RESPONSIVE INNOVATIONS TO DEVELOP AND SPREAD IN THE FACE OF CRISIS. In some cases, this means reducing or eliminating constraints, such as unnecessary requirements or "red tape"; in other cases, it means making public investments in fundamentals such as roads, wholesale markets, and electrification. During the COVID-19 pandemic, the accelerated take-off of e-commerce and the quick spread of alternative payment systems – such as mobile money and informal transfer systems – showed the contribution of such innovations to value chains' resilience. While contract farming and commodity exchanges could also have more important roles to play in price risk management, their growth in LMICs has been hampered by contract enforcement issues, liquidity problems, and high transaction costs, among other issues.[42]

ENSURING CONTINUED PRIVATE TRADING OPPORTUNITIES IS IMPORTANT FOR OVERCOMING CRISES. These opportunities help value chain actors to diversify suppliers and customers and to work around local constraints, including restrictive trade policies implemented in response to crises. Such trade barriers often lead to higher price volatility and higher margins in agricultural markets, and should therefore be avoided. Keeping transportation and appropriate logistics functioning is crucial for maintaining trade in agricultural inputs and outputs, which are typically transported over long distances in both transitional and modern markets. For example, the creation of safe corridors for agrifood products – as seen in the case of the Ukraine war – can sometimes reduce impacts of disturbances.

ATTENTION TO GENDER IS IMPORTANT, AS WOMEN OFTEN SUFFER GREATER SETBACKS FROM CRISES. Ensuring that women retain access to productive opportunities across various nodes in value chains can limit the impact on food security and livelihoods. For example, women must be able to take advantage of digital agriculture and finance innovations as well as training in food safety and other food technology practices. Moreover, women often draw down their savings more quickly than men during crises; to address this, further research is needed on women's coping strategies and ways to improve them.

CAREFUL AND FREQUENT MONITORING, BOTH BEFORE AND DURING CRISES, CAN SUPPORT BETTER TARGETING OF INTERVENTIONS TO CRUCIAL VALUE CHAIN NODES. Given the rapid evolution of crisis situations, updated information on where and when shocks occur and whom they affect is often lacking, including information on road closures, price changes, and product scarcities (see Chapter 2). Before a shock occurs, countries can undertake hazard assessments of their value chains to be better prepared and build resilience in advance. Detailed guidance on such risk assessments for value chains is available.[43] During crises, frequent surveys on important impacts are feasible, given widespread mobile phone use. These phone surveys have been shown to be useful[44] and should be encouraged. For example, the World Food Programme is increasingly using such surveys in crisis situations, although monitoring of value chain agents remains limited.[45] In addition, the increasing availability of big data and improved methodologies to effectively use such data has great potential for better monitoring in these fast-changing situations.

CHAPTER 5

Social Protection

Adaptive Safety Nets for Crisis Recovery

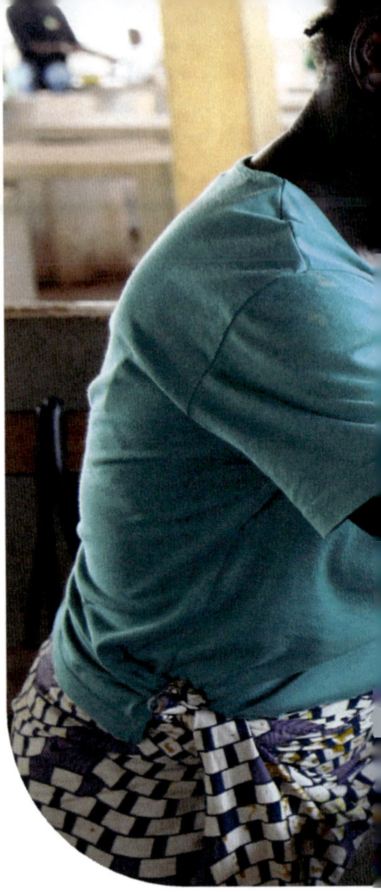

KALLE HIRVONEN

Kalle Hirvonen is a senior research fellow, Development Strategies and Governance Unit, International Food Policy Research Institute.

KEY MESSAGES

- Social protection programs, especially social safety nets that provide cash and in-kind transfers, are an increasingly common policy tool to reduce poverty and improve food security and nutrition in low- and middle-income countries.
- Social protection can play a critical role in times of crisis. Programs have been expanded in response to recent shocks, but coverage remains low in the poorest countries and in urban areas.
- Before crises occur, social safety nets can reduce vulnerability and build resilience by helping households build assets, increase productive investments, and diversify income sources.
- During crises, social safety nets that provide timely and adequate cash or in-kind transfers help maintain household consumption and savings and limit use of welfare-reducing coping strategies. Benefits can be expanded effectively and quickly when programs are already in place.
- There is growing international commitment to better coordinating emergency and long-term social assistance to improve crisis responses.

To boost the role of safety nets in recovery and resilience, steps should be taken to:

- Shift toward a more proactive approach to disasters by building highly adaptive, flexible, inclusive social protection systems and by budgeting for potential crises.
- Invest in incorporating shock-responsive designs into social protection programming to scale up support faster and more effectively during emergencies. This includes investment in monitoring and in predictive early warning systems, as well as unified and digitized targeting systems.
- Improve coordination between emergency humanitarian aid and pre-existing social protection programs to facilitate delivery and targeting of transfers.
- Explore new ways to cover the costs of social protection, such as climate or green financing schemes, and to reduce costs of implementation, such as use of cash transfers and mobile payments when appropriate for the context.

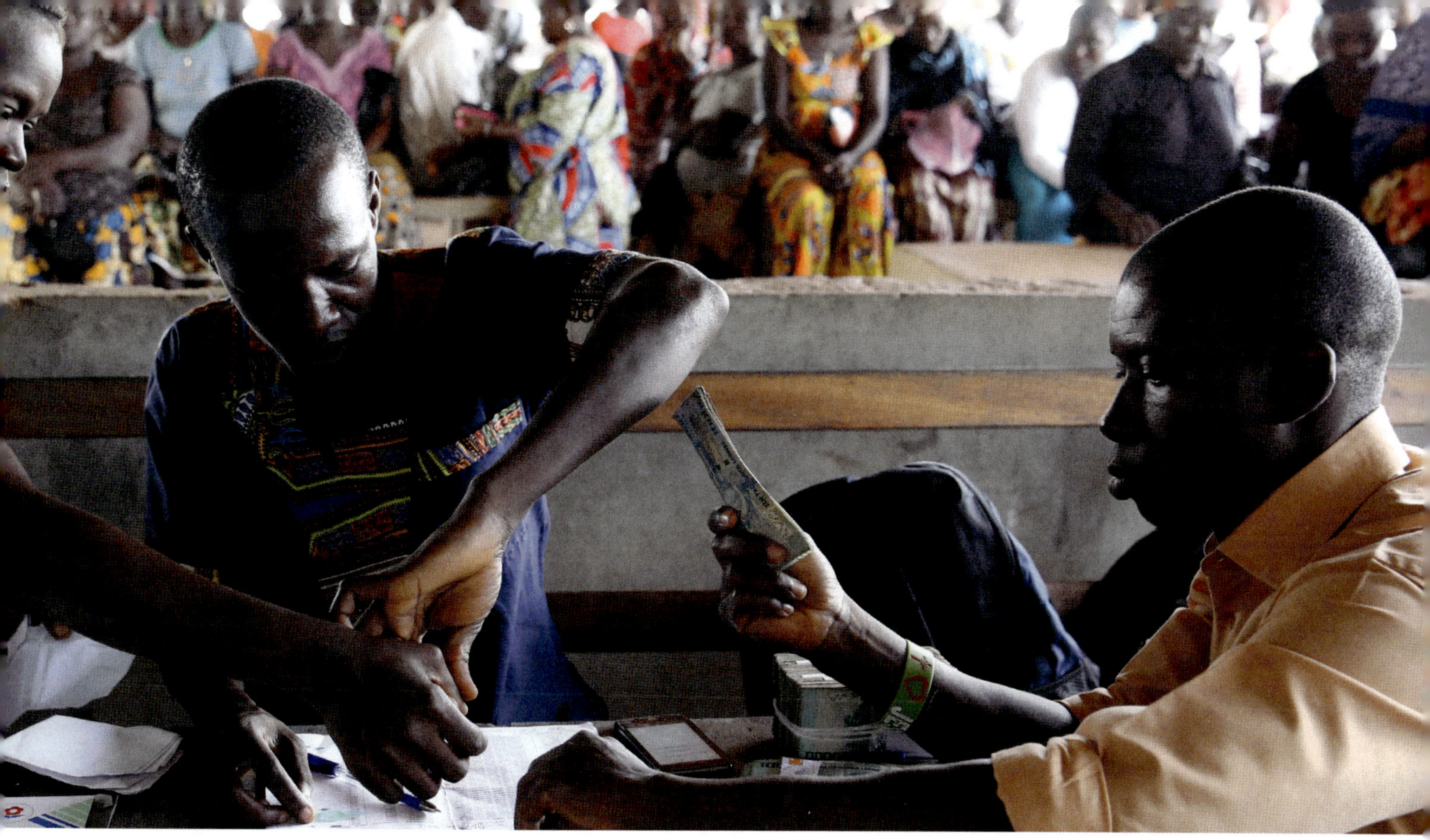

Over the past two decades, social protection programs have become a mainstream policy tool to address chronic poverty and food insecurity in low- and middle-income countries (LMICs). Social safety net programs are one of the most common forms of social protection (Box 1). In sub-Saharan Africa, for example, the number of social safety net programs has more than tripled since the early 2000s,[1] and today each country in the region operates at least one such program.[2] Evidence is mounting that social safety net programs and social protection more broadly can improve food security, reduce chronic poverty, and build household wealth (assets).[3] Moreover, social safety net programs can improve nutritional outcomes,[4] protect aspirations (people's ability to visualize and engage in forward-looking activities) during natural disasters,[5] and increase resilience in the face of climate change.[6] Social safety net programs may even prevent local conflicts,[7] increase trust in local governments,[8] and stimulate economic growth by encouraging savings, addressing credit market imperfections, and creating communal assets.[9] Finally, cash transfers, one form of safety net, have been found to improve women's empowerment[10] and even reduce the risk of intimate partner violence, particularly when coupled with complementary activities.[11]

In low- and middle-income countries (LMICs), social safety net programs reach a considerably larger share of people in rural areas than they do in urban areas (Figures 1A and 1B). Most rural people derive their livelihoods from rainfed agriculture (either directly or indirectly), and therefore many safety net programs have been primarily designed to protect rural livelihoods from extreme weather events, such as droughts and floods. This rural focus is justified, given that global poverty remains concentrated in rural areas[12] and that damaging weather is predicted to intensify and become more frequent due to climate change.[13] However, the COVID-19 pandemic and the 2022 food price spikes – two global shocks – have hit the urban poor particularly hard,[14] exposing the limitations of social protection programming in urban areas.[15] Recurring crises – weather anomalies, natural disasters, disease epidemics, conflicts, and price shocks – are increasingly complex and often

BOX 1 TYPES OF SOCIAL PROTECTION PROGRAMS

Social protection programs fall into three categories: (1) social safety net (or social assistance) programs that provide noncontributory transfers to the poor and vulnerable; (2) contributory social insurance programs; and (3) labor market programs (such as unemployment insurance, wage subsidies, and trainings).[1] Contributory transfers refer to regular payments that individual participants must make to cover the costs of future loss of employment or other shocks. Noncontributory programs do not require payments from the participants.

This chapter focuses primarily on social safety net programs. In low- and middle-income countries, these programs reach a considerably larger share of the population than do social insurance and labor market programs, particularly in the poorest countries (Figure 1). Social assistance programs are also considered more important for poverty reduction than other forms of social protection.[2]

While recent years have seen a shift toward cash-based social protection programming, in-kind transfers in the form of food or nonfood items remain widespread.[3] Transfers can be unconditional or conditioned on recipients meeting certain obligations – for example, education- or health-related objectives, such as participation in classes. In public works programs, transfers are conditioned on work requirements.

Transfers in social assistance programs are typically targeted to the poorest and most vulnerable households. Targeting methods vary. Some programs select beneficiaries based on community assessment or information on household incomes or asset levels while others target certain geographies or demographic groups.[4] Transfers can also be targeted within households, for example to mothers. Recent experimental evidence from cash transfer programs in Burkina Faso, Kenya, and Morocco suggests that whether the targeted recipients are men or women does not significantly affect child health or education outcomes.[5]

Universal basic income schemes provide unconditional transfers to all citizens without targeting. While there have been small-scale pilots in countries such as Finland, India, Kenya, and the Republic of Korea, no country is currently operating a full-scale national universal basic income program.[6]

interlinked, and so require highly adaptive and flexible social protection systems to protect the poor and the vulnerable.

This chapter illustrates how social protection programs protect before, during, and after crises. It then discusses the role of shock-responsive (or adaptive) social protection programs that complement humanitarian response by building resilience before shocks occur and, during crises, by offering a mechanism for channeling support that is both cost-effective and timely.

EFFECTIVENESS OF SOCIAL SAFETY NET PROGRAMS DURING CRISES

Expansion of cash transfers and other social safety net measures has been a common policy response to recent major crises such as the COVID-19 pandemic and the 2022 global food price crisis.[16] In this regard, the 2007/08 global food price crisis was likely an important catalyst, as it alerted policymakers to the inadequate coverage and limited coordination of social protection in LMICs.[17] Ex post assessments found that channeling and targeting of support during the 2007/08 crisis was considerably more effective in countries that had preexisting safety net programs.[18] Possibly as a result, the past two decades have seen major investments in safety nets and other social protection measures in LMICs.[19] Yet despite the strong evidence base and growing interest in expanding social safety net programs, their coverage among the poorest segments of the population remains low in LMICs. According to the latest ASPIRE database,[20] less than 15 percent of the poorest quintile in low-income countries receive social assistance, rising to just below 60 percent in upper-middle-income countries (Figure 1C). Coverage is similarly low for the

FIGURE 1 Share of people receiving different forms of social protection, by country income group

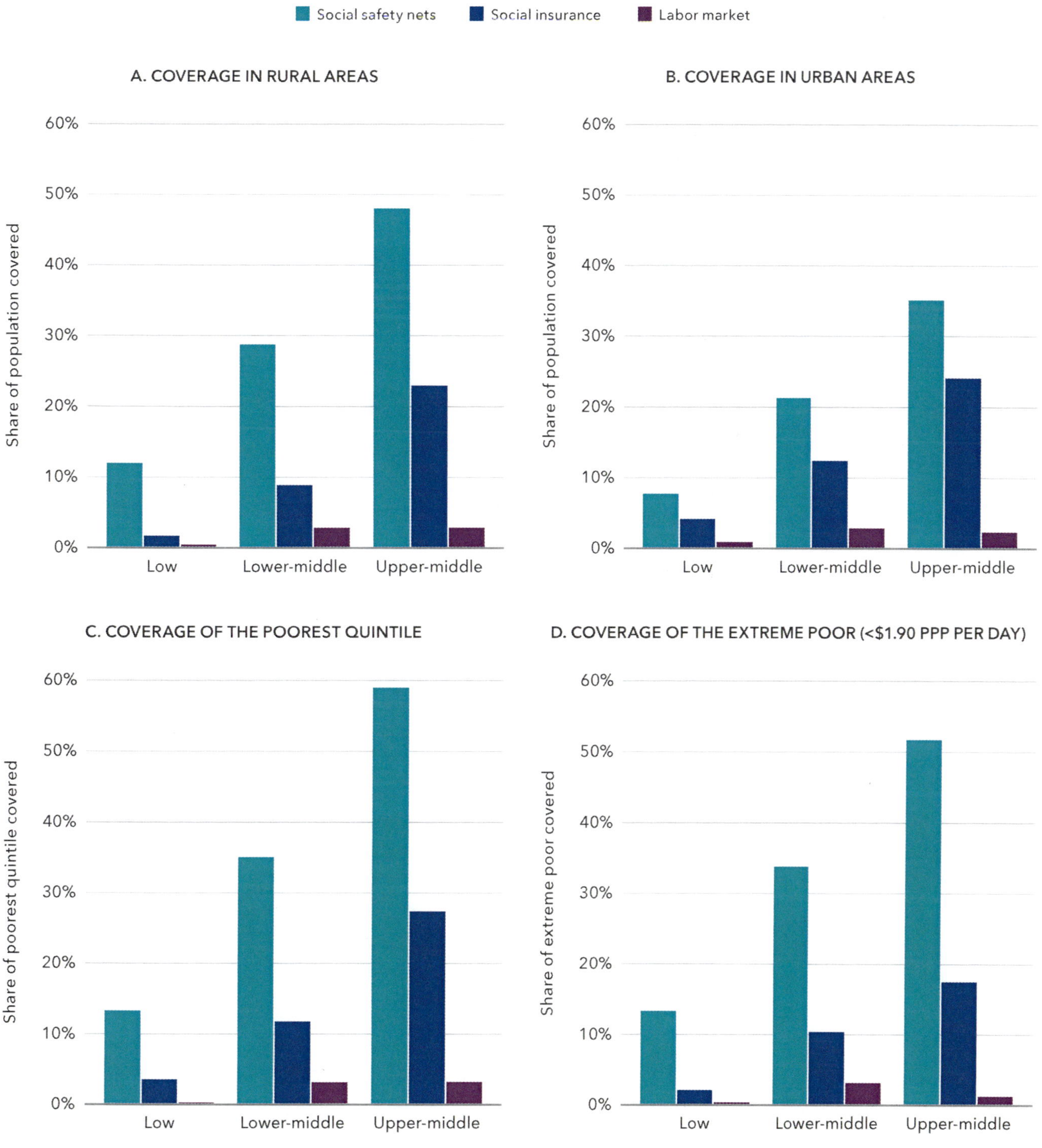

Source: Data from the World Bank Atlas of Social Protection: Indicators of Resilience and Equity (ASPIRE) database (2020).

Note: Social safety nets (social assistance) refers to programs that provide noncontributory transfers to the poor and vulnerable. Social insurance refers to contributory programs requiring regular payments that participants must make to cover the costs of future employment losses or other shocks. Labor market refers to programs such as unemployment insurance, wage subsidies, and trainings. N=112 countries (110 countries in Figure 1D). The latest available year for each country used. High-income countries were omitted due to limited data availability. The poorest quintile (1C) and the extreme poor (1D) are based on per capita pretransfer income or consumption. Missing coverage data were replaced with imputed values using extrapolation or data from the previous available year. If no previous data were available, the coverage level was assumed to be zero.

extreme poor, that is those living with less than $1.90 PPP per day (Figure 1D).

During crises, social safety net programs can offer protection through several channels. Timely and adequate cash or in-kind transfers provide relief in the immediate aftermath of a shock. But safety net programs can also improve resilience by building households' or communities' capacity to deal with future shocks (see Chapter 3). A recent meta-analysis of rigorous impact evaluations found that social assistance programs increase household asset holdings,[21] which can serve as a buffer against future shocks. Safety nets may also promote productive investments and allow households to diversify their income sources, making them less vulnerable to future shocks.[22]

There is growing evidence across LMICs that social safety net programs do protect during crises. In Ethiopia, for example, droughts continue to reduce welfare, but households benefiting from the national Productive Safety Net Program (PSNP) recover to their pre-shock food security levels faster than do nonbeneficiaries.[23] A smaller-scale UNICEF cash transfer program in north Ethiopia was found to protect children's food consumption during localized droughts.[24] Zambia's Child Grant Programme, which provides unconditional cash transfers to households with preschool-age or disabled children, has protected household consumption expenditures during rainfall anomalies.[25] In Niger, an unconditional government cash transfer program mitigated the negative impacts of droughts on household consumption and poverty.[26] Mexico's conditional cash transfer program, Progresa, has been found to protect the consumption of nutritious foods during droughts,[27] keep children in school following natural disasters,[28] and even remedy negative impacts of shocks that occurred several years before households enrolled in the program.[29] In India's Bay of Bengal region, access to a rural livelihood program partly mitigated the devastating economic impacts of an unusually strong cyclone in 2013.[30] In response to a major cyclone in Fiji in 2016, the government provided a one-time top-up transfer to the beneficiaries of existing social protection schemes. Fijian households that were only just eligible for an existing program based on a poverty score index *and* received the top-up transfer recovered faster from the cyclone's damages to their dwellings than households with only slightly better scores that were ineligible to participate in the program.[31]

Disease epidemics constitute a very different type of shock than do weather shocks and other natural disasters. For example, the COVID-19 pandemic resulted in increased mortality and morbidity, but also negatively affected incomes and disrupted food systems, as well as complicating the logistics of delivering assistance through social protection programs.[32] However, evidence from the pandemic suggests that transfer programs also protect beneficiaries during such widespread disease outbreaks. A cash transfer program rolled out in Colombia targeting poor households during the pandemic improved their food access and reduced their reliance on welfare-reducing coping strategies, such as asset depletion and borrowing.[33] In Bolivia, a large-scale noncontributory pension program had sizable positive impacts on food security during the early months of the pandemic, particularly protecting poor households and those who lost their livelihoods.[34] In rural Ethiopia, the PSNP protected household food security during the pandemic.[35] Another approach, a universal basic income scheme in rural Kenya, showed positive effects on food security as well as on physical and mental health.[36] And in urban Kenya, a one-time cash transfer to women-led microenterprises substantially increased inventory spending, revenues, and profits during the pandemic.[37]

These findings from a wide range of contexts provide strong evidence that cash transfers and other social protection measures protect household consumption and savings during natural disasters and epidemics. In the absence of safety nets, poor households usually have no option but to resort to welfare-reducing coping strategies, such as cutting food consumption, selling productive assets, or pulling children from school, with women and girls often the worst affected (see Chapter 6). Such coping strategies can have serious negative consequences in the short term, and their negative impacts may persist for several decades. For example, a sizable literature shows that short-term nutritional deficiencies during early childhood can lower final educational attainment and increase the risk of poverty in adulthood.[38]

SHOCK-RESPONSIVE SOCIAL PROTECTION

Despite the growing number of social protection programs, many LMICs continue to receive emergency aid to address humanitarian situations, many of which are protracted or recurring[39] (see Chapter 7). This fact, and the increasing frequency and complexity of shocks, has generated a widespread commitment among international agencies to strengthen coordination between social protection and emergency aid.[40] Notably, the Grand Bargain agreement between international donors and humanitarian agencies, launched at the World Humanitarian Summit in 2016, commits them to "increase social protection programmes and strengthen national and local systems and coping mechanisms in order to build resilience in fragile contexts."[41] The core premise is that leveraging existing social protection programs as a platform for channeling emergency support can be quicker, more effective, and more inclusive than setting up and operating parallel humanitarian systems during crises.[42] For example, during the COVID-19 pandemic, preexisting social protection programs were often considerably more agile in delivering and targeting transfers than entirely new programs and initiatives.[43]

During crises, emergency aid can be channeled to existing social protection beneficiaries (vertical expansion) or used to expand coverage to crisis-affected nonbeneficiary households (horizontal expansion). Other adaptions include adjusting the rules and conditions of existing social protection programs or aligning the emergency support to match the modalities of an existing social protection program.[44] The past few years have seen an increased interest in establishing such adaptive or shock-responsive social protection programs in LMICs.[45] While rigorous evaluations of these programs are still in the works,[46] many LMICs have already incorporated shock-responsive designs into their social protection programming.

Kenya's Hunger Safety Net Program (HSNP), for example, provides unconditional cash transfers on a bimonthly basis to the poorest households residing in drought-vulnerable northern Kenya.[47] The program is designed to expand horizontally during droughts and other weather shocks. The National Drought Management Authority monitors weather conditions in the area using the remote-sensing-based Vegetation Condition Index (VCI). Very low VCI values trigger horizontal expansion in the form of emergency payments to households not included in the HSNP. The program's budget has been drafted based on needs in normal years as well as careful assessment of drought probabilities and costs of disaster response.[48]

Ethiopia's PSNP was set up to provide a more sustainable response mechanism to recurring droughts and ad hoc emergency appeals in areas that have been historically vulnerable to droughts and other weather disasters.[49] Within these areas, communities themselves select beneficiaries who receive payments for six months, in the form of cash or food, in exchange for performing labor-intensive public works, while poor and chronically food-insecure households with limited labor capacity receive unconditional payments. With 8 million beneficiaries, the PSNP is one of the largest safety net programs in Africa.[50] However, despite its success in improving food security, asset levels, and resilience,[51] the need for annual humanitarian aid persists in areas where the PSNP is operational.[52] At the national level, it is estimated that approximately 5 million people who are not regularly benefiting from the PSNP require emergency assistance in non-drought years,[53] highlighting the chronic gap between actual needs and the funding made available for the program.[54]

The PSNP, however, does have various mechanisms for scaling up support during crises. During a widespread drought in 2011, the program expanded both vertically (by extending the duration of support to 6.5 million beneficiaries) and horizontally (by providing three months of payments to more than 3 million additional people).[55] Leveraging the PSNP during the crisis had multiple benefits. The time between identifying the crisis and responding to it was reduced to two months, compared with the typical response time of eight months for disbursement of emergency support in Ethiopia. In addition, the use of existing delivery platforms was highly cost-effective: an estimated cost of US$53 per beneficiary compared with $169 spent for United Nations or NGO-managed emergency assistance. An evaluation of the coordination

FIGURE 2 Limited association between climate risk and social assistance coverage in the poorest quintile

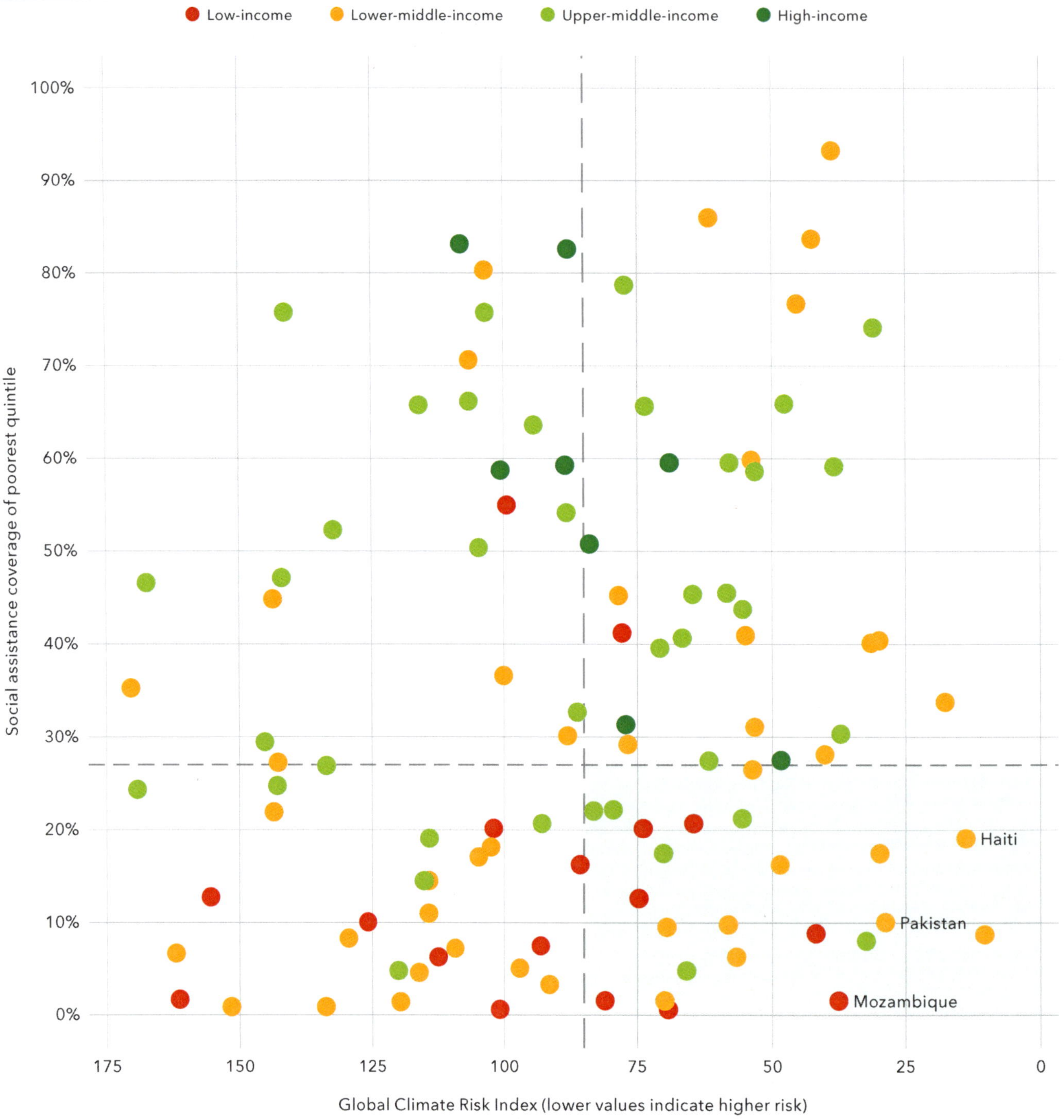

Source: Data from the World Bank Atlas of Social Protection: Indicators of Resilience and Equity (ASPIRE) database, updated June 28, 2022; and from GCIR, D. Eckstein, V. Künzel, L. Schäfer, and M. Winges, *Global Climate Risk Index 2020* (Bonn: Germanwatch, 2019).

Note: The Global Climate Risk Index (GCIR) measures the extent to which countries have already been affected by weather anomalies in terms of lives lost and economic losses. Lower GCRI values indicate higher climate risk. N=120 countries (latest available data point for each country). Dashed vertical and horizontal lines mark the median values of climate risk index and social assistance coverage, respectively. The shaded quadrant indicates the area of greatest concern.

FIGURE 3 Share of ODA allocated to humanitarian aid and social protection

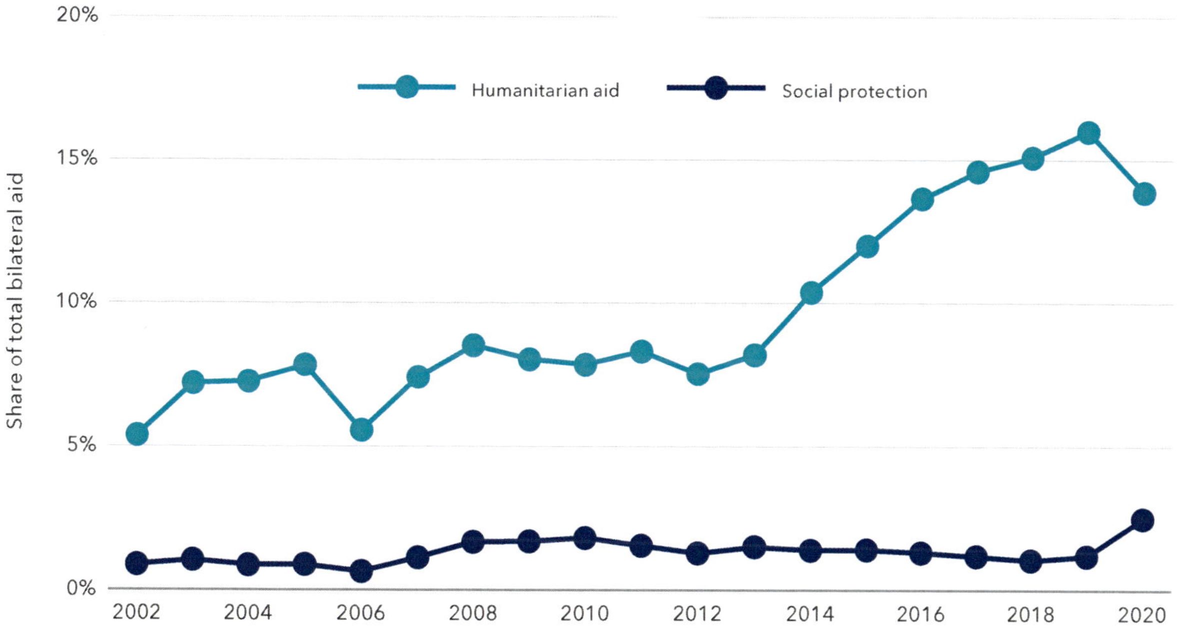

Source: Data from OECD-DAC database, Official Bilateral Commitments (or Gross Disbursements) by Sector: Aid (ODA) by Sector and Donor [DAC5] (Paris: OECD, 2022).

Note: Official development aid (ODA) (from all official donors) disbursements for social protection (ODA category 16010, Social Protection) and humanitarian aid (ODA category 700, Humanitarian Aid, Total) are compared to total ODA disbursements (All Sectors, Total).

between the PSNP and emergency support in 2017/18 found that together these two programs provided a continuum of support: the PSNP targeted chronically food-insecure households, while the humanitarian aid focused more on acutely vulnerable households.[56] Since then, an effort has been underway to further consolidate the PSNP and annual emergency assistance delivery systems and procedures into a single framework.

Setting up shock-responsive social protection programs requires major investment and effort.[57] Effective shock response requires close coordination across different social protection programs as well as emergency response programs within a country. Moreover, the information requirements for these programs are high. Policymakers need to know what risks vulnerable populations are facing, where these risks are likely to materialize, and who is vulnerable.[58] Early warning systems are needed to facilitate a rapid and effective response (see Chapters 2 and 3). In Bangladesh, for example, anticipatory cash transfers to households predicted to be severely affected by impending floods served to mitigate the negative impacts on food security and partially protected household savings when the flooding occurred.[59] Unified targeting systems based on social registries likely need to be established to rapidly determine eligibility for support when crises occur. For example, the introduction of a unified targeting system in Indonesia improved both targeting accuracy and harmonization across different social protection programs in the country.[60] Possibly as a result, Indonesia's social protection response during the COVID-19 pandemic was considered strong: more than 85 percent of households received some form of assistance during the early months of the pandemic and the support was relatively well targeted to the poorest households, with little duplication across different programs.[61]

GOING FORWARD

Some countries explicitly target their national safety net programs to climatically vulnerable areas, characterized by frequent droughts or other erratic

BOX 2 GRADUATION PROGRAMS

Jessica Leight, Research Fellow, International Food Policy Research Institute

In recent years, a growing literature in development economics has examined the complex interrelated constraints faced by households in extreme poverty. Given the salience of these multiple constraints, multifaceted "graduation model" interventions – which simultaneously address several barriers – are widely viewed as promising. The first large-scale evaluation of this approach was conducted as a multicountry trial in Ethiopia, Ghana, Honduras, India, Pakistan, and Peru, analyzing an integrated package of social protection interventions that included two years of consumption-support cash transfers, an asset transfer (valued at between US$500 and $1,000), training, weekly household coaching visits, household-level health training, and savings groups.[1] This package not only led to substantial increases in consumption, food security, assets, and financial inclusion in the medium term, but also its effects persisted 10 years later in India, by which point the consumption impacts had roughly tripled in magnitude.[2] Another large-scale trial of a similar intervention implemented in Bangladesh by BRAC, an international development organization, also showed very substantial positive effects in both the medium and long term, up to 10 years post-intervention, with large increases in consumption, assets, food security, and financial inclusion.[3]

Additional evaluations of graduation programs in conflict-affected areas have been conducted in Afghanistan and Yemen – showing robust positive effects in Afghanistan, but more modest effects only on savings and assets four years post-transfer in Yemen – and in Ghana, where the effects of a more limited set of interventions, including only productive asset transfers or savings schemes, were minimal or zero.[4] A very recent contribution also found that a graduation program incorporating psychosocial support in Niger had positive effects on consumption and food security, income, and mental health in the short term.[5] While the evidence from Ghana suggests that scaled-down sets of interventions including only some of the graduation model components do not have impacts comparable to the full set of interventions, the evidence is nascent and thus this remains an important area for future research.

Overall, major gaps remain in the evidence on longer-term effects and in evaluations of projects implemented at scale or within the context of broader government social protection programs. The original graduation model pilots were generally small in scale. However, the Targeting the Ultra Poor programming run by BRAC in Bangladesh targeted 450,000 households, and the graduation program in Niger was rolled out in the context of a government social safety net, albeit to a subsample of recipient households. Particularly given the high cost and intensive implementation required for graduation model interventions, better understanding of whether they can be effectively scaled up will be a crucial focus for future research.

weather patterns. For example, Niger's adaptive safety net program targets areas exposed to recurrent drought, as determined by an index that considers rainfall and vegetation density data derived from satellite sources.[62] Globally, however, there is only a limited correlation between social assistance coverage in the poorest quintile and the Global Climate Risk Index (Figure 2),[63] which measures the extent to which countries have already been affected by extreme weather events (droughts, floods, heatwaves) in terms of lives lost as well as economic losses. The dashed vertical and horizontal lines in Figure 2 mark the median climate risk and social assistance coverage levels, respectively. The lines divide countries into four quadrants based on their relative level of social assistance coverage and climate risk. The bottom right quadrant captures countries of particularly high concern – countries such as Haiti, Mozambique, and Pakistan are exposed to high climate risk but have very low social assistance coverage for the poorest quintile.

Overall, governments and aid agencies need to shift toward a more proactive approach to disasters, replacing ad hoc humanitarian appeals during crises with social protection programs that build long-term resilience and respond to extreme weather events and other disasters when they occur.[64] The shift should be accompanied by appropriate risk-financing instruments that prepare for disasters before they happen.[65] This entails calculating the odds of disasters occurring in a given region or country and estimating the costs of responding. Budgets can then be drafted not according to the needs in nondisaster years, but at a level that accounts for disaster probabilities and their response costs.[66]

Yet globally, ad hoc responses remain the norm. The share of official development assistance (ODA) allocated to humanitarian aid increased rapidly over the past decade, while the share of ODA allocated for social protection remained relatively stagnant (Figure 3). Considering the solid evidence from a wide variety of contexts showing that social protection programs build resilience and offer protection during crises, thereby reducing the need for humanitarian response, a strong argument can be made for increasing spending for social protection.

Social safety net programs in LMICs depend largely on external funding.[67] To ensure the continuity of these programs, LMICs must diversify funding sources by enhancing domestic revenue collection mechanisms and exploring innovative financing. For example, programs like Ethiopia's PSNP and Indonesia's Keluarga Harapan conditional cash transfer program have been found to increase tree cover or prevent forest loss,[68] thus potentially qualifying them for climate or other green financing schemes.[69] Another way to alleviate the financial burden of social safety nets is to reduce implementation costs. For example, switching from in-kind transfers to cash or mobile payments can produce considerable cost savings for program implementers.[70] However, to minimize harmful effects for transfer recipients, it is important to consider the context before making such adjustments, particularly when food prices are rising rapidly or are volatile.[71]

In the long run, the goal of social safety programs should be to strengthen livelihoods to promote long-term resilience and eventual graduation from assistance. A growing body of evidence shows that carefully designed graduation programs (Box 2) can lift households out of poverty, improve food security, and increase resilience to shocks by unlocking productive investments and permitting households to diversify their income sources.[72]

CHAPTER 6

Gender
Promoting Equality in Fragile and Conflict-Affected Settings

HAZEL MALAPIT AND LYNN BROWN

Hazel Malapit is a senior research coordinator, Poverty, Gender, and Inclusion Unit, International Food Policy Research Institute (IFPRI). **Lynn Brown** is director of Alliances, Policy, and the Africa Region, HarvestPlus, IFPRI.

KEY MESSAGES

- The treatment of women is a better predictor of a state's peacefulness than its level of wealth, status of democracy, or ethnoreligious identity. In fragile and conflict-affected settings, women and girls face disproportionate risks that include forced displacement and gender-based violence.
- Comprehensive and systematic data to provide evidence on the gendered consequences of crises are still lacking, particularly for disasters and conflicts. Yet, sex- and age-disaggregated data are critical to understanding how crises differentially affect women and men, and girls and boys; monitoring whether programs are reaching and benefiting the appropriate groups; and designing gender-responsive interventions.
- Women's voices are rarely heard in disaster management, despite evidence that their participation can improve outcomes, including in conflict situations. Although women are often consulted during the needs assessment phase of response management, they are not involved in the design of projects.

To improve the outcomes of crisis responses, it is important to:

- Prioritize gender targets and track progress, and direct funding toward programming that promotes gender equality and women's empowerment in fragile and conflict-affected settings.
- Adopt innovative methods to address the gender data gap. Providing mobile phone access to women can have multiplier effects, enabling women to receive cash transfers directly while providing a platform for high-frequency data collection and targeted information campaigns.
- Generate more evidence on violence prevention strategies. To date, few studies empirically evaluate the impact of violence prevention and response interventions in fragile and conflict-affected settings, but important research is underway, including work by the interdisciplinary Cash Transfer and Intimate Partner Violence Research Collaborative, hosted by IFPRI.
- Ensure that women's voices are included at all levels, especially in peace processes and in senior management and high-level government positions where policymaking and programming decisions are made.

Gender inequality exists everywhere, but it is particularly stark in fragile and conflict-affected settings (FCAS). Yemen, Afghanistan, Chad, Iraq, Pakistan, and the Central African Republic – all of which are considered extremely fragile and are affected by conflict (except Pakistan)[1] – score the lowest on the Gender Development Index.[2] The States of Fragility 2022 report also finds that the Gender Inequality Index is highly correlated with all six dimensions used to assess fragility, with 15 of 21 conflict-affected contexts facing high levels of gender inequality.[3] The same pattern is shown by the Global Gender Gap Report[4] – 72 percent of the countries ranked in the bottom quartile of gender parity are considered fragile, and a number of countries affected by conflict, such as Haiti, Somalia, and South Sudan, do not even have sufficient data to be included in the ranking.[5]

These patterns indicate a close relationship between fragility and gender inequality. The treatment of women is a better predictor of a state's peacefulness than its level of wealth, status of democracy, or ethnoreligious identity. Democratic states with a higher level of violence against women are as unstable as nondemocratic states.[6] States with male-dominant family law systems have greater levels of fragility,[7] and other quantitative studies show that states with higher levels of gender inequality are more likely to experience conflict.[8] Compared to other developing country contexts, women and girls in FCAS are exposed to greater health risks, such as maternal mortality and early pregnancy, along with other negative outcomes such as forced/child marriage and gender-based violence (GBV).[9] The increased exposure of women and girls to these risks has adverse effects on their human capital, access to resources, and economic participation, which in turn reduces their agency and resilience to manage and cope with other shocks and stressors. Households headed by widows in conflict settings are more vulnerable to intergenerational poverty, and children exposed to related traumas, such as orphanhood, can experience lasting developmental impacts, with adverse consequences for health and economic outcomes in adulthood.[10]

In this chapter, we focus on catastrophic crises – disasters[11] and conflicts – that lead to displacement, which has devastating consequences for women and girls. These types of crises are of particular concern because of the lack of evidence-based policy responses.

GENDERED CONSEQUENCES OF DISASTER AND CONFLICT

Existing evidence confirms that crises disproportionately impact women's assets, livelihoods, and well-being.[12] Such gendered impacts have also been observed from growing evidence on the direct and indirect impacts of current crises, including the COVID-19 pandemic and the Russia-Ukraine war.[13]

When it comes to disasters and conflict, however, the evidence remains quite limited. A decade ago, a review observed that most research on the gendered impacts of conflict focused exclusively on GBV and called for a wider set of issues to be considered.[14] Since 2013, research on GBV in disaster settings and emergencies has grown considerably, although the quality of quantitative studies remains poor.[15] A recent scoping review[16] confirms that, to date, few studies have explored the gendered effects of conflict on agricultural productivity and food security. We also do not know much about the distribution of care work, and gendered impacts in human and physical capital that make long-term movement out of poverty possible. These knowledge gaps limit understanding of the full range of the gendered distribution of impacts, as well as the underlying mechanisms that lead to those impacts – critical information that can help inform both short- and long-term policy responses.

HEALTH AND SANITATION SERVICES AND WOMEN'S WORKLOADS

Disruptions in health and sanitation services in FCAS may exacerbate women's care burdens. In Ethiopia, recent studies on the impacts of the ongoing civil conflict use high-frequency phone surveys to assess food insecurity and access to health and water, sanitation, and hygiene services. IFPRI research found that the outbreak of the conflict increased the probability of moderate to severe food insecurity by 38 percentage points.[17] Additionally, the share of respondents who were unable to access needed health services increased by 35 percentage points, and the share of respondents who were unable to purchase staple foods increased by 26 percentage points.[18] These negative impacts were more pronounced for poor households, rural households, and those with undernourished children.

Although the data do not allow for individual-level analysis and the majority of phone survey respondents were men (around 62 percent across all rounds),[19] it is likely that women's workloads increased disproportionately for households that were unable to access needed health services due to the conflict. According to the 2013 Ethiopia Time Use Survey, women spend more than twice as much time as men on unpaid care work, with 5.5 hours daily for women compared with 2.0 hours for men.[20] The time use data also suggest that prior to the conflict, women already bore a double burden by spending more than an hour longer than men on unpaid and paid work combined, leaving them with less time for rest. The livelihood and income uncertainties accompanying conflict are likely to exacerbate the workload pressure on women, who remain primarily responsible for preparing food and caring for children and other family members. This added pressure also raises the importance of recognizing care needs as part of impact assessments and recovery planning.

WOMEN AND GIRLS FACE INCREASED RISK OF GBV

GBV is a serious public health concern, with nearly one in three women worldwide having been subjected to physical and/or sexual violence.[21] Numerous studies also show that intimate partner violence (IPV) and other forms of violence against women and girls (VAWG) have risen at alarming rates due to the COVID-19 pandemic.[22] Disasters exacerbate stress and violence against women, regardless of country income status. After the 2011 earthquake in Christchurch, New Zealand, for example, IPV reportedly increased by 40 percent in rural areas.[23]

Recent evidence reviews suggest that the risk of GBV is even more elevated in humanitarian and emergency settings, particularly for adolescent girls.[24] The consequences of GBV extend beyond the immediate physical injuries and mental trauma of the experience. Studies show that GBV survivors are more likely to suffer from reproductive health issues, sexually transmitted infections, unwanted pregnancies, depression, and anxiety and to develop unhealthy coping strategies, such as drug use.[25] Moreover, necessary health services to meet the increase in healthcare needs after exposure to GBV and across the life course may be limited or unavailable in disaster and displacement situations.

A recent systematic review examined VAWG in disaster situations across quantitative, qualitative, and mixed-method studies.[26] The authors report that nearly half of all quantitative studies found positive associations between exposure to disaster and some type of violence, and none found that disaster exposure was associated with decreased VAWG. The review uncovered three potential pathways through which disaster exposure can increase VAWG: (1) increased stressors such as poor mental health and loss of housing and livelihoods; (2) poor law enforcement and risky post-disaster housing environments; and (3) underlying drivers that are exacerbated by disaster exposure, such as forced marriage of girls and worsened social norms rooted in men's feelings of inadequacy in the face of disaster.[27]

Results from a systematic review and meta-analysis suggest that one in five refugees or displaced women in complex humanitarian settings[28] experience sexual violence.[29] However, this is likely a significant underestimation, given the well-documented underreporting of VAWG across all settings.[30]

The most prevalent forms of GBV against adolescent girls are child marriage, domestic violence, and sexual violence.[31] Adolescent girls can be separated from their families and support networks during displacement, which contributes to increased risk of GBV. For example, a 1999 government survey in Sierra Leone found that 37 percent of prostitutes were under the age of 15, and of those, 80 percent were children displaced by war.[32]

GIRLS FACE INCREASED RISK OF CHILD MARRIAGE

Marrying in childhood is a human rights violation. It is worth emphasizing that children cannot give informed consent to sex, marriage, or other critical decisions. Early marriage has significant consequences for girls, including curtailment of education, impacts on sexual and reproductive health, exposure to IPV and GBV, early pregnancy and higher maternal mortality and morbidity, sexually transmitted diseases including HIV, and higher rates of under-five mortality for children of girls.[33] These risks can be exacerbated in displacement situations where services are more lacking for adolescents.

Child marriage is more common for girls than boys in FCAS (Figure 1). Marriage rates for girls under the age of 18 exceed 50 percent in Burkina Faso, the Central African Republic, Mali, Mozambique, Niger, and South Sudan. In Niger, almost three out of four girls are married by the age of 18. Additionally, the lasting impact of intermittent fragility and/or frequent disasters on addressing child marriage is shown by child marriage rates in Bangladesh (51 percent), Chad (61 percent), and Guinea (47 percent), which persist even though the World Bank does not classify these countries as fragile due to conflict.

In most of the countries ranked most fragile (marked with [*] in Figure 1), more than 10 percent of girls are married before the age of 15. In the Central African Republic and Niger, more than 25 percent of girls are married before the age of 15.

Poverty contributes to the likelihood of child marriage because marriage can be used by households to reduce the financial burden of caring for or educating daughters.[34] These pressures are likely exacerbated during periods of heightened economic insecurity, but few studies have examined this in fragile settings. For example, a recent report suggests that in Ethiopia, child marriage has more than doubled as families struggle with food insecurity in regions hit hardest by drought.[35] One recent study that tracked whether adolescent girls were more likely to be married as a result of the economic stress from the COVID-19 pandemic found that 18 percent of respondents in Uganda reported knowing of a family in their community who had their underage daughter married due to the pandemic, usually because of parental income loss and economic uncertainty.[36] Nearly half of the respondents reported knowing or hearing about girls in their community becoming pregnant while schools were closed. This highlights the risk of early

FIGURE 1 Child marriage rates (2014–2020) in fragile and conflict-affected countries

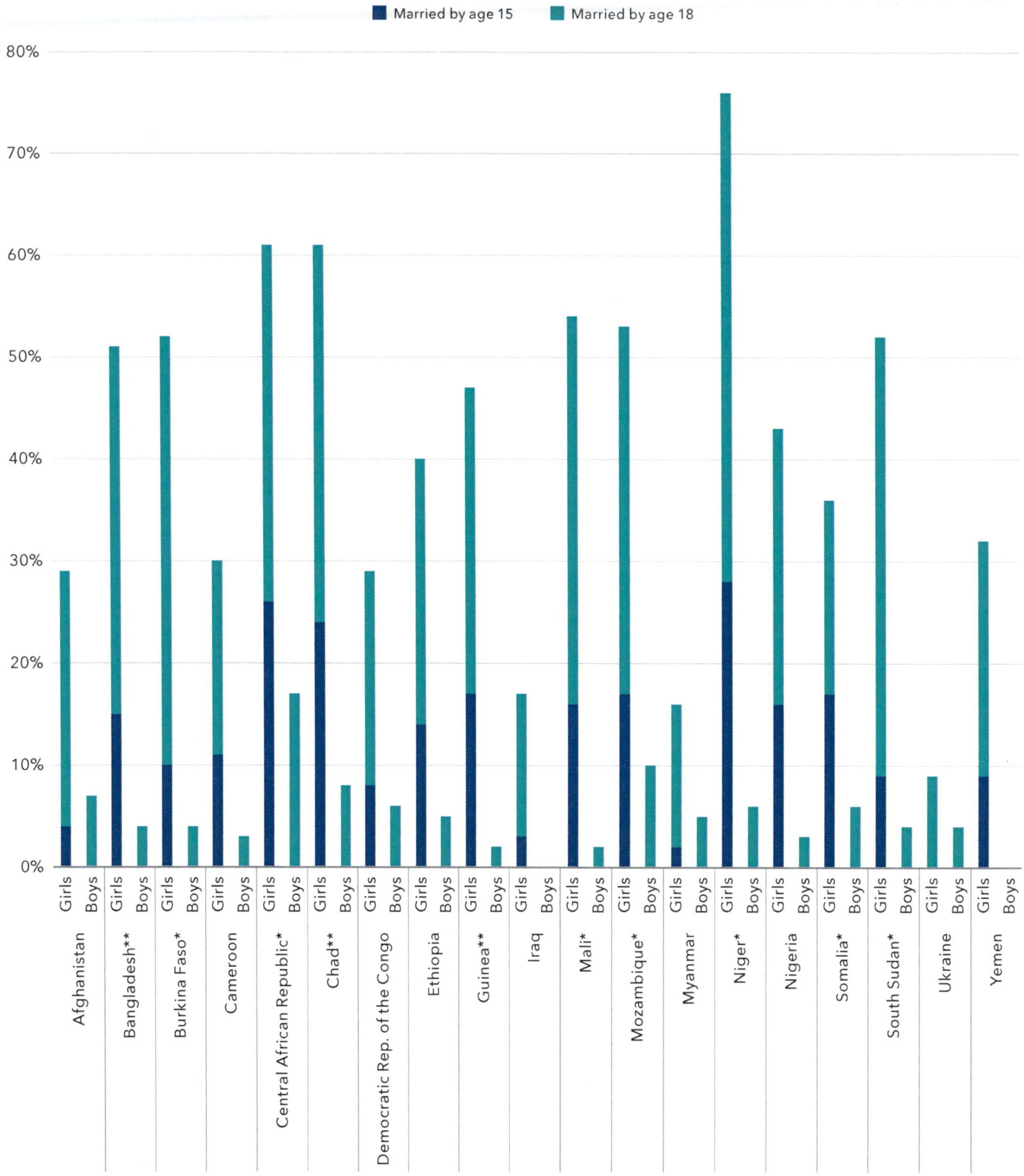

Source: Country list based on the World Bank's FY2023 List of Fragile and Conflict-Affected Situations (2022). Child marriage data are from UNICEF, *The State of the World's Children 2021: On My Mind - Promoting, Protecting and Caring for Children's Mental Health* (2021).

Note: Countries marked with an asterisk (*) are classified as the 10 most fragile, based on the OECD's 2018 *State of Fragility Report*; countries marked with ** are on the OECD list but not the World Bank list.

pregnancy among adolescent girls, which has negative implications for the future health and human capital outcomes of young mothers and their children.[37]

RECOMMENDATIONS

How can programs and policies promote gender equality in fragile and conflict-affected settings?

PRIORITIZE GENDER TARGETS AND TRACK PROGRESS

Programming for gender-specific humanitarian responses has been historically underfunded.[38] Recent commitments to address this shortfall have led to a notable surge in gender-specific humanitarian funding – more than doubling from US$268 million in 2018 to $587 million in 2021 – but this represents only 1.9 percent of total international humanitarian assistance.[39] Moreover, current reporting systems do not allow for accurate tracking of funding commitments, a lack of visibility that increases the risk of funds being diverted away from gender.[40] According to key informants interviewed for the 2022 *Global Humanitarian Assistance Report*, establishing an integrated gender component before the onset of the COVID-19 pandemic minimized the risk of funds being diverted, despite the rising needs from the health emergency.[41] The pressure to deprioritize gender is exacerbated in settings where funding for humanitarian operations is severely limited.

Directing funding toward programming that promotes gender equality and women's empowerment pays off in FCAS. Guidance on the most promising interventions comes from a recent systematic review of 14 intervention types across 29 FCAS.[42] The review finds that most gender-specific and gender-transformative interventions have overall positive effects on the primary dimension of women's empowerment being targeted, and that none of the interventions lead to negative effects on any outcome. Effective gender-focused interventions include cash transfers, self-help groups, village savings and loan associations, and technical and vocational education and training. These can improve multiple dimensions of empowerment, while asset transfers, sensitization campaigns, and capacity-building programs lead to promising results across some dimensions of empowerment. To maximize program effectiveness, the study's authors recommend explicitly targeting specific empowerment outcomes and tailoring the intervention to the drivers of gender inequity in the given context.[43]

CLOSE THE GENDER DATA GAP

Routine collection of sex- and age-disaggregated data is a critical step both to promote gender equality and to support gender-targeted programming in FCAS. Without appropriate and timely data, there is a general lack of knowledge on the gender-differentiated impacts of crises, and policy responses are unlikely to address the most pressing needs of women and men, and boys and girls. A recent Inter-Agency Humanitarian Evaluation found limited evidence of sex- and age-disaggregated data being used to inform the analysis and adaptation of project activities, with negative consequences for the quality of initial response activities for women and girls, as compared to other populations.[44] These policy responses span a wide range of actions that address both the underlying causes of crises (such as development programs to stabilize livelihoods and social protection programs) and the outcomes of crises (such as anticipatory action programming or emergency response humanitarian programming). It is critical that women be counted in policy responses in all these stages – before, during, and after disasters and conflict.

PRACTICAL CHALLENGES

However, data collection during disaster and conflict periods is fraught with practical challenges. Data collection systems in FCAS are rarely set up to systematically collect sex-disaggregated data. Conventional sources of individual-level data, such as household surveys, may be difficult to implement in settings with extensive migration or forced displacement or due to logistical, security, and ethical concerns. Even where available, data may not be interoperable across different systems, limiting how they can be analyzed, and there may be political sensitivities around sharing data across institutions. Other methodological difficulties include the lack of reliable baseline information on

BOX 1 WHEN DISASTER STRIKES, RIGID SECLUSION NORMS CAN MEAN DEATH FOR WOMEN AND GIRLS

Women and girls bear an unequal burden from disasters. According to data from 141 countries affected by natural disasters between 1981 and 2002, disasters lower the life expectancy of women more than of men.[1] Restrictions on women's freedom of movement contribute to their vulnerability during disasters, particularly in contexts where women may not be able to decide whether to evacuate.[2] For example, during the 1993 earthquake in Afghanistan, seclusion norms reportedly prevented women from evacuating.[3] During the 1991 cyclone in Bangladesh, 9 out of 10 deaths were of women.[4] Many women reportedly waited for their husbands to return home before deciding to evacuate, in part due to a lack of warning information, which had been transmitted primarily to men.[5]

Insights from Pakistan reveal how gender norms expose women and girls to death in disaster situations. The 2022 monsoon rain triggered flooding that covered around one-third of the country and left 6 million people in need of assistance. Elders and men in more patriarchal remote villages forbade women from moving to camps where they would be safe from the floods, and would have access to food and water. In Basti Ahmad Din, a small village in Punjab province, its 400 residents faced starvation and disease as the village became an island. More than half of its homes were destroyed by flooding, but the elders forbade women from leaving for relief camps, as it would entail them mixing with men outside their families. Instead, men traveled to the camps to secure supplies for the villagers. In another area of Punjab, similar concerns reportedly led to the death of dozens of women and children. In yet another village, men evacuated to higher ground with their livestock, leaving the women behind.[6]

gender inequalities and other variables of interest during non-crisis periods.[45]

For example, although the number of internally displaced persons (IDPs) more than doubled from 26.4 million to 53.2 million between 2012 and 2021 due to conflict and violence (see Chapter 7), data for people displaced by disaster often reflect only immediate displacement and are rarely disaggregated by sex or age.[46] According to the Internal Displacement Monitoring Centre (IDMC), of the countries and territories from which it collected data in 2018, only 14 percent documented the sex and age of IDPs, and of those, only a quarter did so systematically.[47] In the absence of reliable sex- and age-disaggregated data on IDPs, it is difficult to assess which subpopulations are more likely to be displaced, how long people are displaced for – particularly in disaster situations – or whether they are displaced multiple times in a year.

Similarly, it is unclear how displacement due to disaster affects men and women differently. Do men stay to protect fixed assets in disaster zones and settings affected by extreme weather? How often are women forbidden to move to IDP camps in the face of disaster (Box 1)? Do men return earlier to reestablish homes, leaving spouses and children behind? Are women's assets disposed of first as households rebuild the family home and livelihood during these types of crises? Past research shows that shocks to household livelihoods often result in women's assets being sold first (including productive assets such as small livestock), before those that are controlled and used by men to generate income (such as farm machinery or cropland).[48] These findings suggest that crises can disproportionately erode women's incomes, savings, and assets, which has serious implications for their future livelihoods and bargaining power within the household.

The reality is that neither internal nor external displacement is a short-term occurrence in most instances. The average UN-coordinated humanitarian response to address these situations lasts nine years.[49] The life-course needs of women of child-bearing age and girls can change dramatically within this span of time. Given the lack of sex-disaggregated data, it is unclear how agencies can ensure these needs are addressed, especially

BOX 2 OPERATIONAL RECOMMENDATIONS ON WASTING

The world's most fragile and conflict-affected countries bear the brunt of any food crisis. Wasting is the indicator of choice in these emergency settings, as it changes quickly in response to both food shortfalls and/or disease outbreaks caused by issues with healthcare or access to safe water. In the first six months of 2022, one child became severely wasted every 60 seconds, increasing the total of severely wasted children from 7.674 million to 7.934 million.[1] Severe wasting is a key predictor of child mortality, with mortality rates 11 times higher for severely wasted children than well-nourished ones, and accounts for 20 percent of all global child deaths.[2] Severe wasting that affects more than 30 percent of children under the age of five years in a region is one of three indicators used to declare famine – the others are when 20 percent of the population faces extreme food shortages and when 2 out of 1,000 people die of starvation daily.[3]

Ninety percent of children treated for severe wasting are in emergency settings, including displacement and refugee camps.[4] This is a major challenge for slow-onset disasters, such as severe drought, where parents are unable to protect their homes and assets, and mothers may be very young. Delayed displacement to a camp may result in children dying before arrival or en route, or being too ill to save when they arrive.

Mid-upper arm circumference (MUAC) tapes are a simple, inexpensive tool to check for wasting in children. To do so, the paper measuring tape is placed around the upper arm of a child ages 6–59 months. The tape is color coded, with red indicating severe wasting, amber/orange indicating moderate wasting, and green indicating no wasting. MUAC tapes are commonly used by rural clinics and community health workers. These tapes could be given to women in areas with slow-onset drought or ongoing conflict, which often limits their mobility and access to clinics. By providing these tapes with instructions on their use, women could be made aware that accessing more food or healthcare is critical when the child's arm measurement begins to enter the amber/orange segment of the tape. It would also help women in slow-onset drought disasters, such as in Somalia, leave for displacement camps as a matter of urgency, given that it can take several days of walking to reach these camps.

Mobile phones also offer an opportunity for simple text messaging that encourages women to measure their children and report red or orange measurements, potentially enabling humanitarian workers to identify the most severe problems earlier and mobilize a rapid response.

for young girls who transition to adolescence and face increased threats due to their gender and life stage. This can lead to more child marriage, as parents lack resources and seek to protect their adolescent daughters.

TESTING DIGITAL SOLUTIONS

To this end, closing the digital gender gap can have multiplier effects. If every displaced woman entering an IDP camp were provided a mobile phone as part of the humanitarian assistance package, then the scope would widen considerably for both data collection and gender-responsive interventions. This idea is being tested on the ground by the World Food Programme, which is scaling up the use of mobile money for humanitarian assistance and prioritizing women as recipients of food assistance and cash transfers.[50] Apart from mobile money transfers, the phone could also be used to collect high-frequency data on women and children and provide low-cost interventions, such as targeted nutrition messaging. For example, if women received both a mobile phone and a mid-upper arm circumference (MUAC) tape as part of their humanitarian assistance package, they could receive instructions and periodic prompts via phone to use the tape to monitor their children's wasting status, which is a critical predictor of child mortality, particularly in emergency response settings (Box 2). Independent access to a mobile phone can also help deter GBV by making it easier for women to report sexual harassment and other violations to trusted authorities. Husbands may still control if and when women are allowed to use mobile phones in some contexts, however, so it would be important to understand what conditions make this type of intervention more likely to succeed.

Experience from the Gender, Climate, and Nutrition Integration Initiative phone surveys suggests that it is possible to collect survey data on different aspects of well-being, as well as data related to agency, decision-making, and even more sensitive topics such as intrahousehold conflict and gendered practices including child marriage. Because of the sensitivity of some questions, it is recommended that only one respondent be selected in each household (either a man or woman) to minimize the potential of intrahousehold conflict, and that speakerphone use be checked to ensure women's privacy.[51]

GENERATE MORE EVIDENCE ON VIOLENCE PREVENTION STRATEGIES

Despite the increased attention to GBV, few studies empirically evaluate the impact of GBV prevention and response interventions in disaster and conflict settings. Many widely accepted strategies for preventing and responding to GBV in humanitarian settings do not easily lend themselves to experimental designs, including good practices around case management and referral systems, justice and legal aid, safety and risk mitigation, and coordination, assessment, monitoring, and evaluation.[52]

Nevertheless, important research is underway in this field. The latest Sexual Violence Research Initiative Forum, held in 2022, featured new research from humanitarian and displacement settings in Iraq, northern Uganda, South Sudan, and southern Lebanon.[53] These studies examine the integration of violence prevention strategies with livelihood and economic programming. However, some methodological challenges remain, particularly around identifying causal impacts and distinguishing between the impacts of economic components and violence prevention components.

Recognizing that economic insecurity is a well-known risk factor for multiple forms of violence against women and children, the interdisciplinary Cash Transfer and Intimate Partner Violence Research Collaborative hosted by IFPRI aims to build evidence on how cash transfer programming can catalyze IPV prevention among poor and vulnerable women in low- and middle-income settings. With a diverse portfolio of eight studies to be completed in seven countries by 2024, the Collaborative is expected to contribute to a new wave of research that aims to go beyond demonstrating whether cash transfers reduce IPV to explore how practitioners can maximize impacts and whether these impacts can be sustained.[54] Insights from this research will undoubtedly offer lessons for FCAS, where economic distress is a commonly cited contributor to GBV.[55]

LET WOMEN LEAD

Of the 130 peace agreements signed between 1990 and 2014, only 13 included women signatories.[56] Compared to peace agreements without women signatories, those signed by women have not only been more durable, but have also included a larger number of agreement provisions and led to a higher rate of provision implementation 10 years after signing. Enabling women's voice in peace negotiations is associated with a 35 percent increase in the probablity that an agreement lasts at least 15 years.[57] Even when women are not signatories to peace agreements, their engagement in negotiations increases the likelihood of an agreement being signed. Women's influence is often stronger for more fundamental reforms, including postconflict female political representation and legal reforms related to land ownership, inheritance, GBV, and healthcare. In Liberia, women's political activism against violence was critical to ending the country's 14-year civil war.[58] Liberian women continued their advocacy in the aftermath of the Accra Peace Agreement, which led to the historic presidential election of Ellen Johnson Sirleaf, the first female head of state in Africa.[59]

In disaster management, however, women's perspectives are rarely considered.[60] Although the humanitarian system has shown improvements in women's representation in senior leadership,[61] this does not necessarily translate to real influence in response management and programming. Women are often consulted during the needs assessment phase, but they are not involved in the actual design of projects. Their inputs are often limited to hygiene or sexual and reproductive health, rather than their other broader needs, strengths, resilience, and capacities.[62]

The global community should learn from and invest in grassroots women's groups that are leading programs to respond to crises and rebuild livelihoods in their own communities. Women's groups can provide a platform for collective action by sharing labor and childcare responsibilities, organizing transport, accessing credit and savings, and disseminating information.[63] For example, during the pandemic, the Self-Employed Women's Association in India served as an intermediary between female farmers and the government, helping women to sign up for government relief and organizing members to sell their vegetables.[64] Other examples abound, such as women's organizations in Albania, Brazil, Ethiopia, Lebanon, Nepal, and Paraguay that are supported by UN Women.[65] Women's groups know their communities best and can reach those who are most in need. Beyond more financial support, women deserve a seat at the table to shape the policies and programs that directly impact their own lives and communities.

CHAPTER 7

Forced Migration
Fragility, Resilience, and Policy Responses

MANUEL A. HERNANDEZ, OLIVIER ECKER, PETER LÄDERACH, AND JEAN-FRANÇOIS MAYSTADT

Manuel A. Hernandez is a senior research fellow, Markets, Institutions, and Trade Unit, International Food Policy Research Institute (IFPRI). **Olivier Ecker** is a senior research fellow, Foresight and Policy Modeling Unit, IFPRI. **Peter Läderach** is CGIAR co-lead of climate security, The Alliance of Bioversity and CIAT. **Jean-François Maystadt** is a professor of economics, Université Catholique de Louvain and Lancaster University.

KEY MESSAGES

- The decision to migrate is complex, driven by a wide range of context-specific push and pull factors, including economic, social and cultural, environmental, and safety factors.
- Forced displacement – when people must leave their original place of residence – results from various triggering factors, events, and shocks. These include climate change, armed conflict, criminal violence, and economic shocks, which are often interrelated, multiplying their impact. About four-fifths of displaced people have experienced acute hunger and malnutrition.
- Migration, including forced migration, constitutes an important adaptation strategy, with both challenges and opportunities. It can have benefits for migrants and for hosting and sending communities. It is a fundamental component of economic development, allowing individuals to respond to economic incentives or seek out better opportunities.
- Policies that restrict the rights of migrants to work and choose a place of residence in hosting countries should be considered barriers to economic and social integration and development.
- Migration requires resources and socioeconomic networks, and often those who stay behind are the most vulnerable.

To improve the outcomes of forced migration, it is critical to:

- Invest in research to develop better-tailored policies that expand the positive effects of migration and limit negative ones on migrants and their families, sending communities, and hosting communities.
- Adopt nontraditional methods and analytical approaches to trace migration. These can provide new research avenues to better understand the key factors driving forced migration, including irregular migration, which is inherently more difficult to measure and analyze.
- Align social protection and climate action objectives. As conflict and climate change further worsen the global humanitarian crisis – and drive forced migration – humanitarian and climate investments must mutually support peace, security, and climate adaptation and mitigation.
- Identify measures for accelerating the transition from humanitarian aid to development policy and for better integrating refugees into hosting communities. Different options should be considered for integration, with special attention given to the needs of displaced women.
- Prioritize addressing "forced immobility" (that is, the situation of those who are not able or choose not to relocate) – a problem that has received little policy attention.

Migration is a recurrent, complex, and multidimensional phenomenon driven by a broad set of factors. These include both "push" factors that encourage or force people to move from their current location and "pull" factors that attract people to a new location.[1] Migration is also an important adaptation strategy and development pathway that can support livelihoods, build resilience, and protect against fragility and armed conflict. Natural barriers and policy restrictions to migration may similarly result in important welfare losses.[2]

Individuals or households migrate for multiple reasons, including being forced to leave their homes due to climate change, armed conflict, criminal violence, or economic needs, among other triggering factors. This chapter discusses migration as a result of "forced displacement," which occurs when people must leave their "original place of residence as a result of an idiosyncratic shock, whether manmade or environmental."[3] Interactions among these driving forces, such as conflict and/or extreme weather events combined with food insecurity, may also lead to threat-multiplying effects.[4] Recent examples of forced migration include refugees[5] displaced by the Syrian civil war and by the Russia-Ukraine war, the Rohingya people fleeing violence inflicted by Myanmar's state forces, Venezuelan migrants seeking asylum to escape food insecurity and oppression, and people from Central America taking treacherous routes to the United States to escape gang violence and persistent poverty.

Of people forcibly displaced worldwide, as of mid-2021, 80 percent had experienced acute food insecurity and high levels of malnutrition.[6] The COVID-19 pandemic also increased the vulnerability of displaced people and migrants. In East Africa, including the Horn of Africa, for example, the challenges of displaced people were exacerbated by reduced humanitarian funding, a decrease in remittance flows due to travel freezes, and hundreds of thousands of job losses.[7]

Even migration forced by war and violence requires resources and relies heavily on networks.[8] People with more liquid resources are more able to flee,[9] though perhaps less likely to do so,[10] while better social networks can also facilitate

FIGURE 1 Key migration facts

ONE IN EVERY SEVEN PEOPLE IN THE WORLD IS A MIGRANT

763 million are internal migrants and 281 million are international migrants.

INTERNATIONAL MIGRATION SURGED BY 107 MILLION OVER THE PAST 20 YEARS

52% of international migrants are men, about one-third are 15–34 years old, and a large share originate from rural areas.

THERE ARE ABOUT 84 MILLION INTERNALLY DISPLACED PERSONS, REFUGEES, AND ASYLUM SEEKERS

Most people displaced by armed conflict or other forces are from developing countries, and 80% experience acute food insecurity.

CLIMATE DISPLACEMENT HAS RECEIVED SPECIAL ATTENTION IN RECENT YEARS

75% of recent displacements are due to natural disasters, and many people displaced by climate change are women, who are also at greater risk of violence.

FORCED MIGRATION MAY ALSO RESULT IN IRREGULAR MIGRATION

Apprehensions at the US-Mexico border set a new record in fiscal year 2022 and almost tripled compared to 2019.

Source: Data from FAO, *Migration, Agriculture and Rural Development* (Rome: 2016); IOM, *World Migration Report 2020* (Geneva: 2020); J. Barchfield,"Pandemic Deepens Hunger for Displaced People the World Over," UNHCR, March 31, 2021; USAID, U.S. Government Global Food Security Strategy (Washington, DC: 2021); IOM, "Migration in the World," and "Key Migration Terms," accessed January 2023; OHCHR, "Climate Change Exacerbates Violence against Women and Girls," (2022); USCBP, "U.S. Border Patrol Apprehensions," Washington, DC, Dec. 19, 2022.

migration.[11] Thus, migrants are not necessarily those most affected by triggering factors, and they may be relatively better equipped with knowledge and skills that are useful for adaptation in hosting communities than those who remain behind.

RELEVANT MIGRATION FACTS

Worldwide, one in every seven people is a migrant, whether a forced or voluntary migrant[12] (Figure 1). Of these approximately 1 billion migrants, 763 million are estimated to be internal migrants (migrating within their country of origin), while 281 million are international migrants. International migration has received more attention recently, as it surged by 107 million between 2000 and 2020. During this period, Western Europe and the United States were the main destinations for migrants. Among international migrants, 52 percent are men and roughly one-third are between 15 and 34 years of age. About 40 percent of international remittances are sent to rural areas, reflecting the rural origins of many migrants.[13]

While the increase in international migrants has mostly occurred in high-income countries over the past three decades, the rising refugee population has been concentrated more in low- and middle-income countries (Figure 2). The number of refugees has roughly doubled since the early 2000s, reaching 27 million in 2021, and more than 86 percent of them have been hosted by these countries.

Forced displacement may also result in irregular migration, which is the movement of people that occurs outside of the laws and regulations of the sending, transit, and receiving countries.[14] Irregular migration is generally more difficult to track, and there is more information on irregular migration flows to Europe and the United States than within Africa, Asia, and Latin America, where they are likely to be significant.[15] Although stringent border controls and migration policies at the onset of the COVID-19 pandemic led to a temporary decrease in irregular migration, these crossings seem to have resumed – and even increased – since 2021.[16] Apprehensions at the US-Mexico border, for example, numbered 800,000 in fiscal year (FY) 2019,

FIGURE 2 International migrants and refugees in low-, middle-, and high-income countries

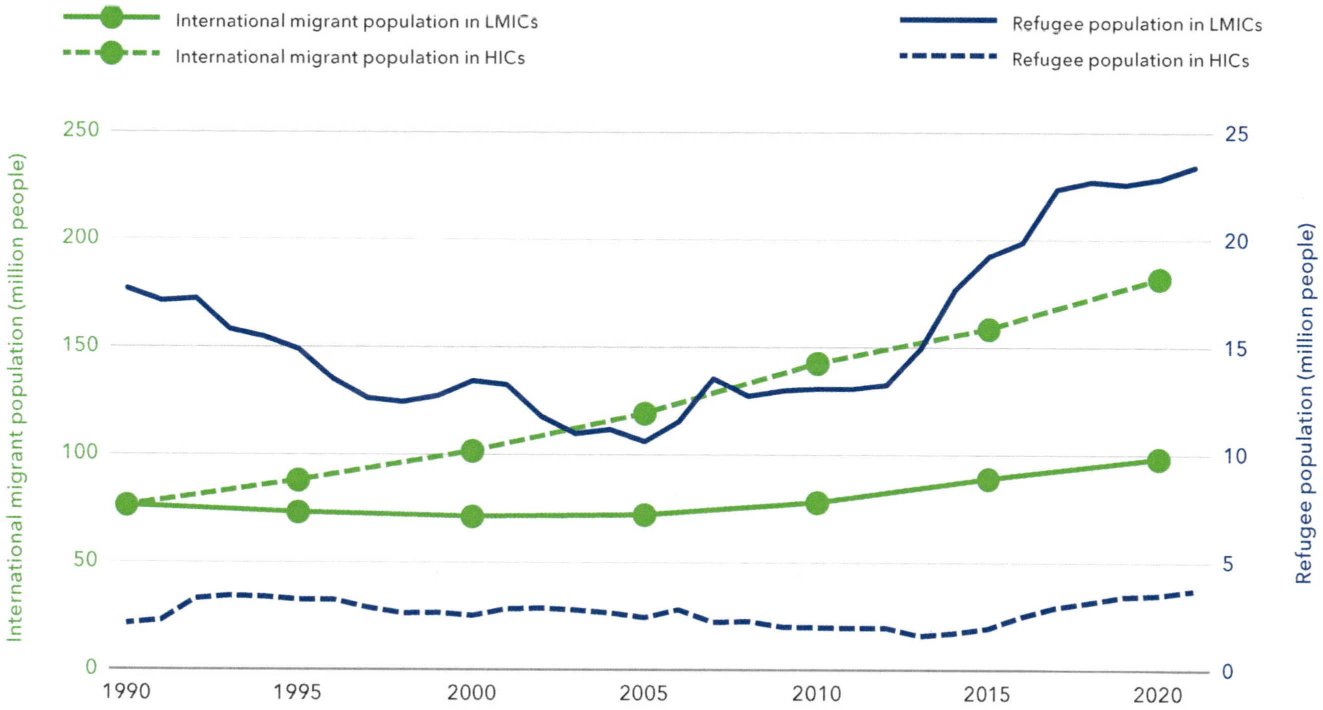

Source: UNHCR Refugee Data Finder. https://www.unhcr.org/refugee-statistics/

Note: LMICs = low- and middle-income countries; HICs = high-income countries.

400,000 in FY 2020, more than 1.5 million in FY 2021, and 2.2 million in FY 2022.[17]

CAUSES OF MIGRATION

Formally identifying migration-triggering factors requires a careful and comprehensive analysis, as many of the factors that influence migration decisions are interrelated, vary over time, reinforce one another, and cannot always be observed.[18] Factors that drive migration are generally grouped into four categories: environmental (such as extreme weather events), safety (such as political instability, conflict violence, and crime), economic (such as income shocks or job opportunities), and social/cultural (such as family and social networks) (Figure 3). These triggering factors are also context-specific and may vary by region or country. They can occur at the individual or household level, as well as the local, regional, and national levels.

Most often, a combination of factors triggers migration. The decision to migrate may be associated with climatic conditions and extreme events; conflict, violence, and crime; food insecurity and malnutrition; job opportunities (or lack thereof); social and political instability in the local area; and/or illegitimate institutions and government repression, among others. In the case of forced internal migration in Africa and the Middle East, for example, the main driving forces include conflict and insecurity, repressive governance, lack of economic opportunities, and climate shocks.[19] The major drivers of irregular migration from Central America to the United States include unemployment (especially among youth), transnational ties (family networks), victimization (crime), and agricultural stress due to natural disasters[20] (although most of the available studies are based on anecdotal evidence and cross-sectional assessments that only allow us to approximate correlations rather than causality).

While Europe is currently experiencing its largest refugee crisis since World War II – with close to 8 million people fleeing Russia's war on Ukraine,

FIGURE 3 Factors driving migration

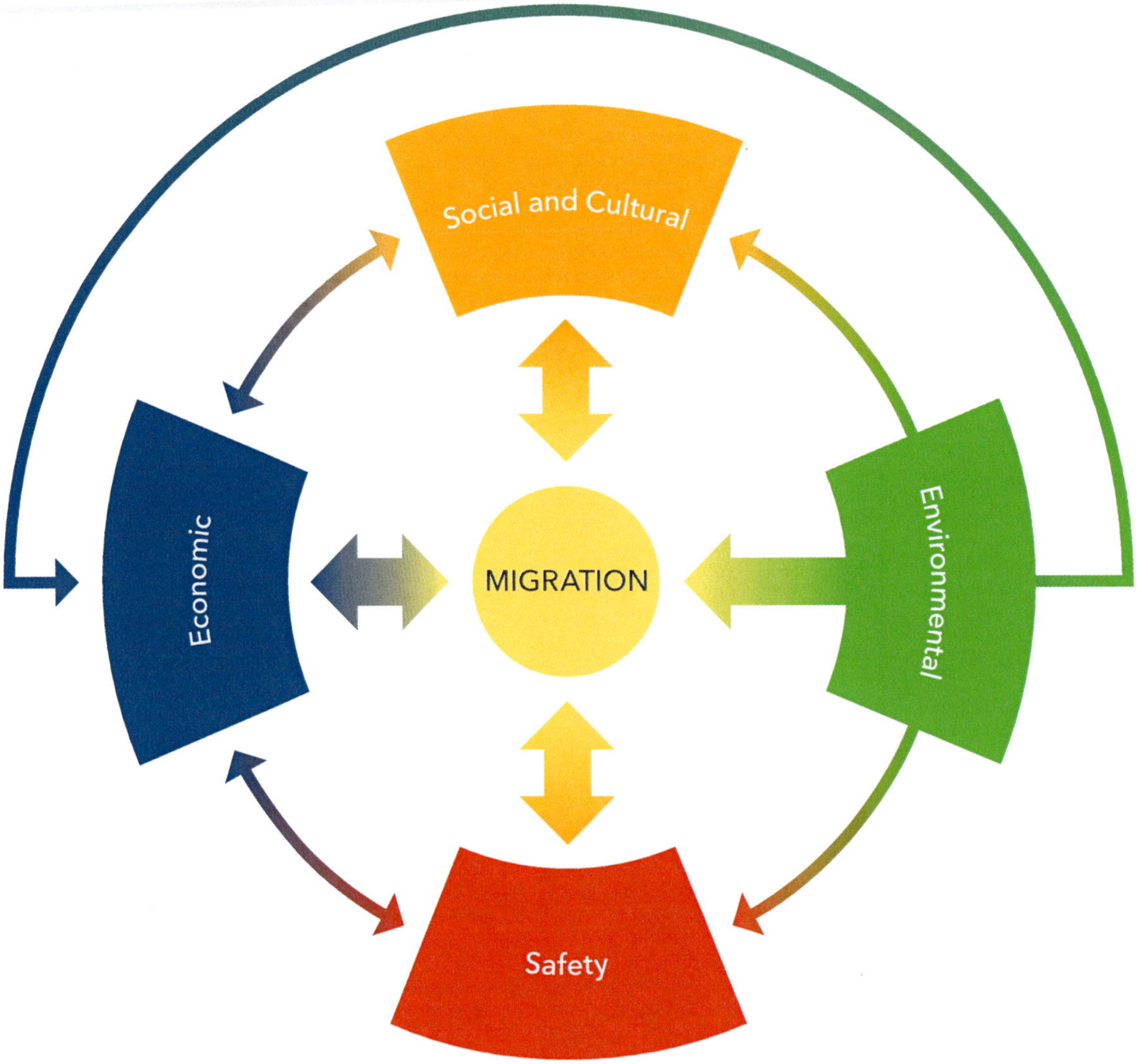

Source: Authors' own elaboration.

Note: The arrows indicate the direction of causality, which is bidirectional in most of the cases between each factor and migration, as well as between the factors themselves.

the vast majority of people displaced from their homes by armed conflict or other forces globally are from developing countries.[21] As of 2019, the International Organization for Migration (IOM) reported more than 84 million internally displaced persons (IDPs), refugees, and asylum seekers.[22] Three-quarters of all IDPs (34.5 million) were living in 10 countries, with half of them in Syria, Colombia, and the Democratic Republic of the Congo. Of all new internal displacements in 2019, 25 percent were triggered by conflict violence and 75 percent by natural disasters. Similarly, of the estimated 26 million refugees worldwide in 2019, two-thirds were from 5 countries (in order of refugee population: Syria, Venezuela, Afghanistan, South Sudan, and Myanmar).

In recent years, special attention has been paid to climate displacement, which occurs when migration is driven, at least in part, by the impacts of climate change. In 2016, the United

Nations General Assembly adopted the New York Declaration for Refugees and Migrants, which explicitly recognizes that people move "in response to the adverse effects of climate change, natural disasters (some of which may be linked to climate change), or other environmental factors."[23] Climate change has been linked to an increase in migratory movements that result from attempts to adapt to the changing environment.[24] Research has also shown that, on average, people move from countries of higher vulnerability to lower vulnerability.[25] This is consistent with the idea that migration is an adaptation to climate change[26] or a response to natural disasters, where families and social networks among migrants in the destination country can play an important (host) role in response to shocks in their country of origin.[27] Areas severely affected by climate change are also more prone to conflict.[28] According to UN Environment, an important share of people displaced by climate change are women, who are also at greater risk of violence, including sexual violence (see Chapter 6).[29]

Recent studies highlight the varying profiles of migrants who are forced to leave their communities and the different reasons driving their decisions. A synthesis brief from the CGIAR Research Program on Policies, Institutions, and Markets[30] provides several key findings from recent CGIAR work on migration drivers:

- The factors driving migration, whether forced or voluntary, generally differ between men and women, and by age. Men are more often motivated by employment, while women face higher barriers to employment and migrate for marriage or educational opportunities.[31] Although both men and women may migrate in response to an income shock, men are more likely to do so.[32] Youth migration is associated with lack of access to land and pursuit of education, although migration does not always lead to more education.[33]

- Climate-driven migration varies by region and country and may differ by age, sex, and socioeconomic group.[34] Adaptation to climate change may reduce migration,[35] while conflict may lead to migration (though this is not always the case).[36]

- Social protection programs have different effects on migration for men and women. For women, these programs may decrease migration, while effects for men may also depend on other factors, such as weather or socioeconomic status.[37] In addition, migrant networks can play an important role, particularly for permanent migration where job search costs tend to be higher.[38]

SOCIOECONOMIC CONSEQUENCES OF FORCED MIGRATION

The consequences of migration are diverse and should be analyzed across three dimensions: impacts on migrants and their families; on sending communities; and on hosting communities.

For migrants themselves, migration may eventually lead to higher incomes and improved livelihoods in the hosting country, including better education and nutrition outcomes for their children.[39] However, these positive effects are not without costs and can take time to materialize, leaving migrants in vulnerable positions that include lower job quality than local workers and deteriorated physical and mental health and well-being.[40] IDPs are more vulnerable as they are more difficult to locate and tend to receive less international assistance. Moreover, migration generally occurs at great risk, with many migrants undergoing extreme hardship and even losing their lives in the journey.[41] According to the Missing Migrants Project of IOM, more than 50,000 people have lost their lives during migratory movements since 2014. More than half of these deaths occurred en route to and within Europe, and around 5,000 people have died or disappeared en route to the United States.[42]

For the families who stay behind, remittances from migrants can constitute an important source of income, allowing them to invest more in education and housing and to attain a better quality of life.[43] Remittances were especially important as a source of income during the COVID-19 pandemic. In 2021 in Latin America, they accounted for 28 percent of GDP in Honduras, 27 percent in El Salvador, 18 percent in Guatemala, 16 percent in Nicaragua, and 4 percent in Mexico.[44] In the Pacific Region, these shares were even higher: 44 percent in Tonga, 32 percent in Samoa, 12 percent in the Marshall Islands, and 9 percent in the Philippines and Fiji.[45] Despite these benefits, migration may also result in an increased work burden for family members who stay behind.

In sending communities, migration may put more pressure on wages for unskilled agricultural workers, which can have serious consequences for the farmers who hire them.[46] Migration may also affect women's workloads and empowerment,[47] and women do not necessarily benefit from the "feminization" of agriculture – that is, the increase in women's labor in agriculture, in their labor relative to that of men, or in their roles in agricultural decision-making (see Chapter 6).[48] Lastly, migration may result in either a "brain drain" or "brain gain" for sending communities.[49] High returns on human capital (education and skills) in the destination country can lead to high-skilled emigration but may also encourage nonmigrants to invest in human capital.[50]

For hosting communities, the economic literature assessing the effects of forced migration is growing, although still limited.[51] Research focused on the African context showed that forced migration is not an economic burden for hosting communities, at least not in a lasting way.[52] On the contrary, these migrants tend to contribute positively to local economic growth. In Rwanda, for instance, each additional refugee has been estimated to increase annual real income in the local economy by US$205 to $253 through market interactions between refugees and their hosts.[53]

Nevertheless, findings also point to rather strong distributional effects for hosting communities, especially in the short term. In the context of underdeveloped labor and credit markets, the poor – who are most vulnerable to livelihood shocks – face the greatest challenges in seizing new economic opportunities that accompany inflows of forced migrants, due to their low levels of physical and human capital.[54] Intrahousehold distributional effects have also been identified, where women with low levels of education are less likely to engage in employment outside of the household.[55] The evidence from African countries is consistent with the findings of more recent studies in the Middle East – the destination of most Syrian refugees – and Latin America – the destination of many Venezuelan refugees.[56]

More recently, researchers have started to investigate whether migrants, and especially those fleeing armed conflict, are more inclined to engage in criminal activities and organized crime in hosting countries.[57] The limited evidence from a few middle-income countries provides mixed and inconclusive results,[58] which emphasizes the need to better understand group dynamics among migrants and intergroup attitudes in refugee camps and hosting communities. The claim that cross-border refugee flows are responsible for propagating localized armed conflict has been stubbornly persistent, especially in the context of civil conflict in Africa – though it lacks strong supporting evidence. A recent study reexamining the effects of refugees on civil conflict found no evidence that hosting refugees raises the likelihood of new conflict, prolongs existing conflict, or increases the number of violent events or casualties.[59]

RECOMMENDED POLICY RESPONSES TO FORCED MIGRATION

RECOGNIZE MIGRATION AS A MULTIDIMENSIONAL, COMPLEX, AND CONTEXT-SPECIFIC PHENOMENON. Policy responses should start from a clear understanding of the causes of forced migration, which may be context-specific, and of the people who migrate, as well as the possible consequences for migrants and their families, sending communities, and hosting communities. A comprehensive analysis is required to determine key driving forces that push (or pull) people to relocate, which often interrelate or intersect in complex ways depending on each setting. New analytical approaches, such as machine learning, and unconventional data sources, such as geo-localized cell phone records or geotags posted to social media, provide new opportunities to fill gaps in data and knowledge about private migration decisions,[60] including irregular migration, which is inherently difficult to trace. Results using these data should still be interpreted cautiously because of likely biases in reporting and selection (the most vulnerable may not have access to tracked communication technology). Although humanitarian assistance is essential in the short term to prevent hunger, malnutrition, and disease among migrants, lasting solutions require wide-ranging policy strategies. These may be tailored to different situations to address the structural causes of forced migration, including lack of economic opportunities, food insecurity, and inadequate access to basic services, and to mitigate the impacts among migrants as well as sending and hosting communities.

BROADEN THE SCOPE OF RESEARCH ON MIGRATION DECISIONS AND POTENTIAL IMPACTS. More research is needed to better understand migration decisions and their potential effects (beyond short-term impacts) on migrants, sending communities, and hosting communities to derive more tailored policies that expand positive effects and attenuate negative ones. For instance, despite a rapidly growing literature examining the socioeconomic impacts of forced migration among host populations in developing countries, surprisingly little is known about the impacts among the migrants themselves or about the costs of "forced immobility" for those who are not able or choose not to relocate. Although evidence is still limited, cash-based transfers or vouchers to refugees have shown efficiency in improving food security among refugees in Kenya, Rwanda, and Ecuador (see Chapter 5).[61] Addressing forced immobility should also be a policy priority. In contrast, cash transfers (for example, cash-for-work programs) in sending communities may increase (rather than deter) migration by alleviating liquidity and risk constraints and not necessarily increasing the opportunity cost of migration (that is, potential gains of staying) among likely migrants.[62]

ALIGN SOCIAL PROTECTION AND CLIMATE ACTION OBJECTIVES. The climate crisis is exacerbating many underlying drivers of conflict and threatens to worsen the humanitarian crises, with ever more people living in fragile and conflict-affected settings. Climate adaptation, peace, and social protection objectives need to be well aligned, especially considering that funds are typically insufficient to cope with multiple crises. Climate investments should be used to support peace, security, and social protection in addition to climate adaptation and mitigation, while humanitarian investments need to support climate action in addition to social protection schemes.[63] In Colombia, for example, a project led by the International Center for Tropical Agriculture (CIAT) is implementing sustainable land use systems to contribute to forest conservation, climate protection, and the peacebuilding process (Box 1).[64]

PROVIDE OPTIONS TO MITIGATE MASS MIGRATION RISKS. Research has been limitied on the potential consequences of different policy options to mitigate detrimental impacts associated with large migration flows in hosting communities. Exceptions include studies focusing on the benefits of local initiatives to better integrate forced migrants into hosting communities, on Uganda's social protection programs for refugees, and on Colombia's right-to-work policy for refugees.[65] Yet beyond these insightful case studies, systematic evidence is still lacking on how specific policies toward forced migrants may lead to improved development and better integration of these populations into their hosting communities.

BETTER TAILOR REFUGEE-TARGETED INTERVENTIONS TO INCREASE THEIR EFFECTIVENESS. Most studies focus on refugees living in camps, while globally most refugees in developing countries live outside of camps.[66] Particular attention should be given to displaced women, given their likely vulnerability to domestic and other forms of violence, the disruption in their access to critical services and informal safety nets, and their lower employment opportunities (see Chapter 6). Geographic mobility has been found to be key for integrating forced migrants in high-income countries, but little is known on the pros and cons of allowing such mobility in developing countries. Migrants respond to economic incentives, and migration itself can lead to a more efficient allocation of resources.[67] From a policy perspective, it is important to consider different options for the reception of forced migrants. Refugees should be allowed to move to local labor markets that offer favorable employment opportunities. Providing them with the option of choosing where to relocate could result in the most effective allocation process.

BOX 1 THE IMPORTANCE OF AMNESTY FOR REFUGEES IN COLOMBIA

Since 2017, more than 5.1 million Venezuelans have fled their country due to its collapsing economy, political turmoil, and humanitarian crisis. Two million of these refugees have relocated to Colombia, although the lack of resources in the hosting country has resulted in a need for long-term solutions and initiatives to promote the socioeconomic recovery of refugees. While previous studies have primarily focused on cash transfers and their effects on refugee welfare, little is still known about the impact of large-scale amnesty initiatives to regularize migratory status and work permits, particularly in developing countries, which often face structural problems such as discrimination in the labor market.

A recent study assesses the impact of the Permiso Especial de Permanencia (PEP) program in Colombia, which has allowed more than 442,000 refugees to find formal employment and access safety nets by regularizing their status. The study shows improvements in several outcomes, such as formal employment rates, poverty levels, access to financial services, per capita income and consumption, food security, and physical and mental health, among those who received the PEP (compared to nonrecipients). These findings demonstrate the importance of a well-conducted amnesty program to smoothly integrate migrants into their hosting communities and improve their well-being.

Source: A. Ibáñez, A. Moya, M.A. Ortega, S.V. Rozo, and M.J. Urbina, "Life Out of the Shadows: Impacts of Amnesties in the Lives of Refugees," Policy Research Working Paper 9928, World Bank, Washington, DC, 2022.

Similarly, offering them opportunities to enroll in training programs that prepare them to actively participate in local labor markets and increase their language skills can enhance their employment prospects in the hosting community.[68]

PROVIDE INCLUSIVE INTERVENTIONS FOR COMMUNITIES HOSTING REFUGEES. More work is needed to understand the impact of refugee-targeted interventions on host communities. In some contexts, for example, cash transfers for refugees can have a large positive impact on food consumption without affecting prices, while in others they may contribute to inflation and resentment toward the refugee population.[69] Refugees may also influence local politics by altering the support for certain parties or affecting voting behavior,[70] which can have important implications for local development. Providing aid and developing infrastructure in the hosting community, including improved public service delivery, can prevent tensions between refugees and locals. More generally, assessing the potential economic burdens of a massive influx of migrants on local infrastructure and social services can help to promote better policies for inclusion.[71]

Overall, forced migration is a recurrent phenomenon that should be incorporated into the global development agenda, given its magnitude and importance for economic development, as it reflects multiple challenges and opportunities for vulnerable populations. It is imperative to invest in more research to better understand migration causes and consequences, including context-specific factors, and to derive better-tailored policies that comprehensively address the phenomenon in both sending and hosting communities.

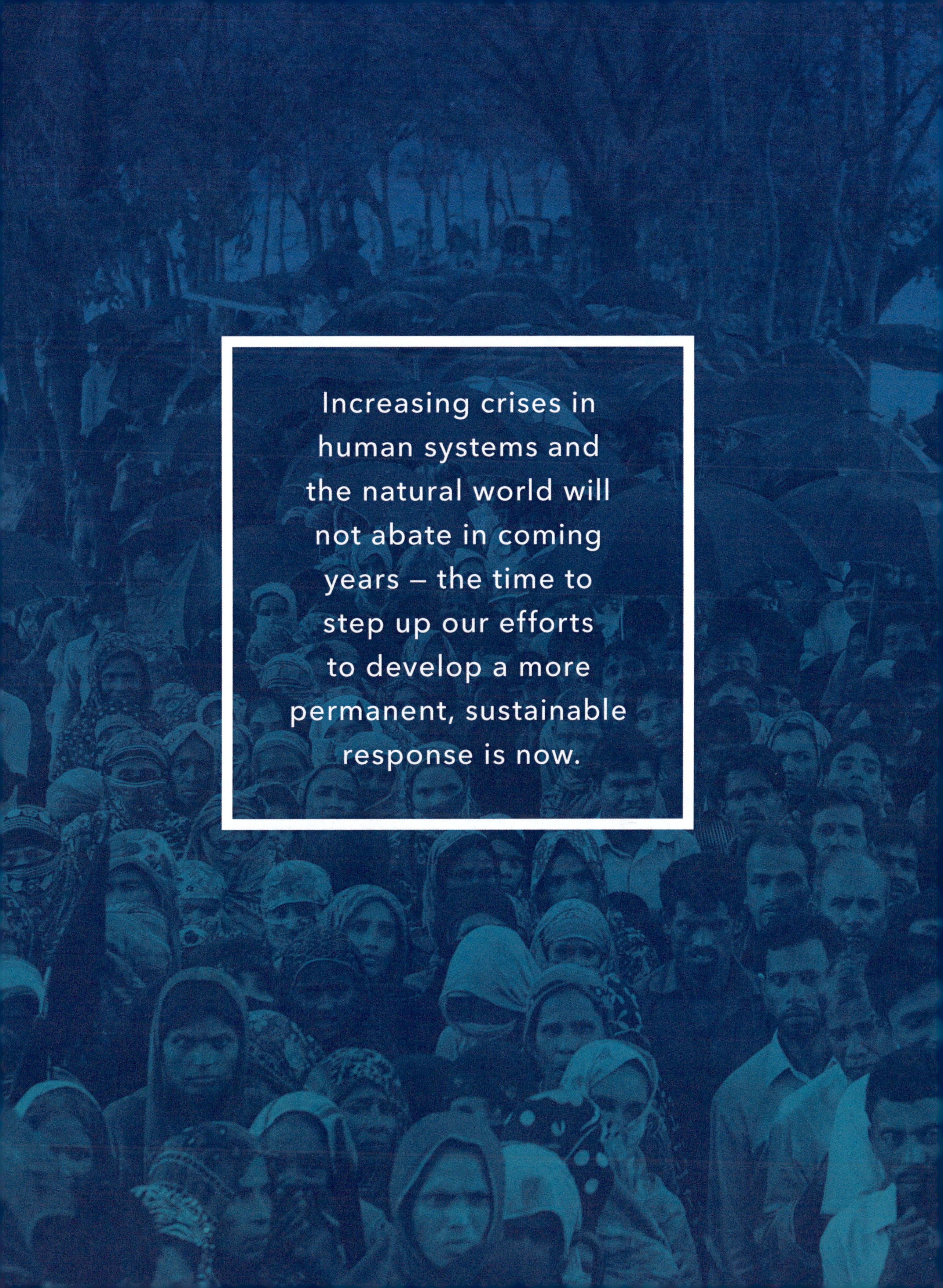

Increasing crises in human systems and the natural world will not abate in coming years – the time to step up our efforts to develop a more permanent, sustainable response is now.

REGIONAL DEVELOPMENTS

RECENT GLOBAL CRISES HAVE LED TO DIVERSE IMPACTS ACROSS THE WORLD'S low- and middle-income regions, reflecting local conditions and differing policy responses. These effects are often compounded by more local shocks and crises, including prolonged conflict and violence, natural disasters, and fragile economic and governance systems. This section examines the impacts of recent food crises to identify both future risks and promising policy options that could improve early warning, immediate response, and resilience building in each region.

- Pursuing a humanitarian-development-peace approach to Africa's protracted crises
- Reducing reliance on food imports in the Middle East and North Africa
- Diversifying trade and improving governance for great resilience in Central Asia
- Increasing smallholder productivity and sustainability in South Asia
- Building regional integration in East and Southeast Asia to better manage future crises
- Managing commodity cycles and building human capital in Latin America and the Caribbean

AFRICA

SAMUEL BENIN, WIM MARIVOET, HARRIET MAWIA, AND JOHN ULIMWENGU

Samuel Benin is deputy division director, **Wim Marivoet** is a research fellow, **Harriet Mawia** is a research officer, and **John Ulimwengu** is a senior research fellow, all with the Africa Regional Office, International Food Policy Research Institute.

In Africa, about 282 million people (20 percent of the population) are facing food insecurity and are undernourished, more than double the share in any other region of the world.[1] Food insecurity levels vary significantly across and within Africa's subregions. As of 2021, countries in central and southern Africa had the largest populations deemed at crisis levels or worse of food insecurity (45.6 million people, 18.4 percent of the population), with 9.9 million at an emergency level (Table 1; see Chapter 2, Box 2, for a definition of the IPC food insecurity phases).[2] In eastern Africa, about 43.6 million people (9.8 percent of the population) are in crisis or worse, with 10.1 million in emergency. In western Africa and the Sahel region, 30.4 million people (8.6 percent of the population) are in crisis or worse, about 42 percent of them in Nigeria.

In terms of absolute numbers of people, the situation is most critical in the Democratic Republic of the Congo (DRC), where 27.3 million people are in crisis or worse, followed by Nigeria and Sudan. In terms of the share of population, South Sudan is most affected, with 60 percent of the population (7.2 million), including 2.4 million people in emergency and 100,000 in catastrophe situations.[3] Other countries with more than 30 percent of the population in crisis or worse include Angola, the Central African Republic, Eswatini, Ethiopia, Lesotho, Madagascar, Namibia, and Zimbabwe.

DRIVERS OF FOOD CRISES IN AFRICA

Food crises in Africa are driven largely by conflict, weather shocks (especially droughts and floods), and poverty, all of which affect the demand, supply, and availability of food.[4] Food shortages and income losses have been worsened by pests associated with extreme weather, especially the fall armyworm plague that started in 2016 in western Africa[5] and the locust infestation across eastern Africa in 2020.[6]

Agricultural policies have also contributed to persistent food crises. Policy support tends to favor agricultural exports, for which prices have been declining, over food commodities consumed in Africa, for which prices have been increasing. Lower export prices have led to declining foreign exchange receipts and income losses, while rising food prices have resulted in higher food import bills and declining investment in agriculture and other key public goods and services.[7]

Other recent shocks compounding food insecurity include the Ebola outbreaks in western Africa (2014–2016) and the DRC (2018–2020), the COVID-19 pandemic, and the Russia-Ukraine war. During the Ebola and COVID-19 outbreaks, lockdowns implemented to limit the spread of disease in many countries led to a slowdown or shutdown of economic activities that disrupted food systems.[8] The continuing crisis reflects remaining supply chain issues caused by the pandemic, as well as additional disruptions from the Russia-Ukraine war,

TABLE 1 Acute food insecurity in sub-Saharan Africa regions and selected countries (millions of people affected), 2021

Region/country	Number of countries included	Integrated Food Security Phase Classification (IPC)			
		Phase 2: Stressed	Phase 3: Crisis	Phase 4: Emergency	Phase 5: Catastrophe
Central and Southern Africa	12	67.7	35.7	9.9	0.01
Democratic Republic of the Congo		40.8	20.5	6.7	0.0
Mozambique		8.4	2.6	0.3	0.01
Malawi		6.3	2.5	0.1	0.0
East Africa	9	51.2	30.3	10.6	0.5
Ethiopia		17.2	12.1	4.3	0.4
Sudan		16.5	7.1	2.7	0.0
South Sudan		3.3	4.7	2.4	0.1
West Africa and the Sahel	16	74.3	28.4	1.5	0.0
Nigeria		35.0	12.7	0.2	0.0
Cameroon		5.8	2.4	0.3	0.0
Niger		5.8	2.4	0.1	0.0

Source: Data from FSIN and GNAFC, *2022 Global Report on Food Crises* (Rome: 2022).

Note: See Chapter 2, Box 2, for information on the IPC classifications.

as a large share of Africa's food imports (especially wheat and maize) come from Russia and Ukraine.[9]

The incidence and severity of these shocks, as well as the drivers, vary across the continent (Table 2). While occasional conflict occurs in many places, several African countries – including Nigeria, Ethiopia, the DRC, Somalia, Mali, and Burkina Faso (in order of fatalities) – suffered substantial violence against civilians in 2022.[10] Conflict, political instability, and violence against civilians are the primary drivers of food crises in other countries as well. The impact of weather shocks is likewise varied and widespread. In 2022, for example, floods affected millions of people and their livelihoods, destroyed thousands of homes and properties, and killed nearly 2,000 people, while desertification and drought are the main challenges in other places.

Poverty has also put healthy diets out of reach for many Africans. Although the cost of a healthy diet in Africa (US$3.46 per person per day) is slightly below the global average (US$3.54 per person per day), per capita income is also lower and poverty rates are higher in Africa than the global average. As a result, a larger proportion of Africa's population cannot afford a healthy diet, especially given recent shocks that have raised food, fertilizer, and fuel prices.[11] The continent's population growth, at about 2.5 percent per year compared to the global average of a little under 1 percent per year, puts additional pressure on the food system and economy to keep pace.

GENDERED EFFECTS OF FOOD CRISES

Food crises affect women and men and boys and girls differently due to norms and cultural practices that lead to different roles, responsibilities, and access to resources and coping strategies (see Chapter 6). Data from several African countries indicate that more women (32.8 percent) than men (29.7 percent) were significantly affected by food price shocks during the COVID-19 pandemic, because women spend a much higher share of their income on food. Moreover, women face hunger more often than men during food crises; for example, in 2014–2016, 25.2 percent of African women were severely food insecure compared to 23.7 percent of men.[12] This disparity is due to differences in income, access to employment or means of production, and cultural practices that put women last, or allot them smaller portions, when

TABLE 2 Main drivers of food crises in selected African countries

Country	Main drivers of food crises
Burkina Faso	Coup d'état in September 2022 and the presence of armed groups, mainly in the country's north.
Chad	Desertification, including drying up of rivers and lakes in recent years, accelerated by drought in northern Chad.
Democratic Republic of the Congo	Combination of increased food prices and transportation costs, epidemics, and one of the world's longest-running armed conflicts.
Ethiopia	Civil war (November 2020 to November 2022) exacerbates the effects of drought.
Kenya	Multiple shocks including dry spells, below-average crop and livestock production, localized resource-based conflict, and the COVID-19 pandemic.
Malawi	Poor infrastructure keeps vital aid from reaching the poorest parts of the country.
Mozambique	In Cabo Delgado province, extremist groups have forced more than 700,000 civilians from their homes since 2017.
Niger	In 2021, a surge in armed groups and internal conflicts forced tens of thousands of vulnerable people into the driest parts of Niger.
Nigeria	Loss of more than 860,000 acres of land every year to desertification, affecting 11 of 36 states.
South Sudan	Decades of armed conflicts, including eruption of civil war in 2013, frequent climate-related shocks (severe flooding and dry spells), and macroeconomic crisis.
Uganda	Drought in 2022 led to price increases of up to 25 percent for basic household items.
Zimbabwe	The 2018/19 drought plus long-standing macroeconomic challenges are pushing millions to the edge of starvation.

Source: Authors' compilation based on Convoy of Hope, "Food Crisis in Africa Reaches Terrifying Levels," Aug. 25, 2022; IPC, "Acute Food Insecurity and Malnutrition Snapshot Acute Food Insecurity: October 2022 - July 2023, Acute Malnutrition July 2022-June 2023" (2022); République Démocratique du Congo, "Aperçu de la sécurité alimentaire et de la nutrition, juillet 2022-juin 2023" (2022).

food is in short supply.[13] In Sierra Leone and Liberia, for example, the closure of food and other retail markets to control the 2014-2016 Ebola outbreak destroyed the livelihoods of traders, 85 percent of whom were women.[14] Similarly, in South Africa, women accounted for about two-thirds of the job losses during the COVID-19 lockdowns.[15]

Such disruptions can exacerbate other negative impacts for women and girls, such as violence and sexually transmitted infections. For example, sexual and domestic violence reportedly rose in Ebola-affected regions of the DRC after an outbreak began in 2018.[16] Likewise during the Ebola outbreak in Guinea, a 4.5 percent increase in violence against women was reported.[17] Food insecurity can also increase the likelihood that women and girls will engage in negative coping strategies, such as transactional sex, to generate income needed to purchase food for their families.[18] Conflict seems to widen the gender gap as well (see Chapter 7). Some studies have found higher rates of chronic malnutrition among pregnant women and children or increased risk of acute malnutrition in areas of several African countries affected by armed conflict, including Burundi,[19] Côte d'Ivoire,[20] Ethiopia and Eritrea,[21] Nigeria,[22] Rwanda,[23] and Somalia.[24]

CRISIS RESPONSES AND CHALLENGES

National and international actors (such as governments, UN agencies, and NGOs) as well as affected local communities and households have responded to food crises with varied approaches and coping strategies.

HUMANITARIAN ASSISTANCE is the most common, straightforward response to aid affected populations. In 2022, the total budget for the UN's Humanitarian Response Plan for sub-Saharan Africa was estimated at US$16.7 billion. This funding is largely earmarked to ensuring food security, while a smaller amount is allocated to nutrition, refugees, and social protection. However, as of the end of October 2022, less than 45 percent of total humanitarian needs had been funded (see Chapter 3).

EARLY WARNING SYSTEMS have emerged as a critical instrument to increase the effectiveness and efficiency of humanitarian responses over the years (see Chapters 2 and 3). Studies show that projections for Africa from famine early warning systems, such as the Famine Early Warning Systems Network (FEWS NET), are generally good, but sometimes miss the mark. These forecasting issues are usually associated with complex climate and weather events, as well as the difficulty of predicting the impact of conflict on food insecurity, as conflict-affected areas are hard to access and politically sensitive to analyze (see Chapter 3).[25] To facilitate early action, some early warning systems and emergency preparedness initiatives, such as the work of the Africa Centres for Disease Control and Prevention, have integrated surveillance and response strategies to mitigate the impact of disease outbreaks.[26] However, like other early warning systems, these too face challenges with data and information management systems, laboratory capacity and functionality, and human capacity, especially in the most remote areas.[27]

MIGRATION is another common response to food crises, and can take many forms depending on where migrants go, the duration of migration, and recurrence. Each choice is driven by a particular set of pull and push factors, and leads to diverse outcomes for migrants and the sending and host communities (see Chapter 7).[28] The total number of intra-African migrants increased from about 13 million people in 2000 to more than 20 million in 2020, with internally displaced people (IDPs) fleeing conflict and violence accounting for most of the increase. Displacement may also be triggered by climate change and extreme weather events, such as the flooding in 2020 that affected more than 2 million people across 18 western and central African countries.

RESILIENCE BUILDING has gained traction over the past decade as a potentially cost-effective strategy to tackle underlying vulnerabilities and spur local solutions for highly contextual challenges.[29] This strategy focuses on creating and rehabilitating household and community assets, including strengthening institutions to manage their ownership, access, and use. In 2021, for example, the World Food Programme reached 2.1 million people across 12 western African countries through its Food Assistance for Assets program. This program, which was gradually introduced beginning in 2013, has assisted local communities in restoring or cultivating 75,000 ha of agricultural land and constructing or rehabilitating 1,400 km of water infrastructure and 244 km of feeder roads.[30] The protection and restoration of ecosystems that provide essential services can be an important component of resilience building (Box 1). However, assessing the impact of any resilience-building intervention is difficult given the multiple definitions of and metrics on resilience, the complex nature of the intervention packages, the difficulty of tracking intervention costs, and the uncertain timeframe for recovery.[31]

THE HUMANITARIAN-DEVELOPMENT-PEACE (HDP) NEXUS APPROACH aims to strengthen collaboration, coherence, and complementarity among these three pillars of crisis recovery (see Chapter 7). Given that any external intervention may have significant consequences – both intended and unintended – on local power balances, institutions, and social cohesion, the HDP approach works to ensure that interventions maximize the reduction of vulnerability and poverty while addressing the root causes of conflict.[32] One good example is the Partnership for Recovery and Resilience, which was set up in South Sudan in 2018 and has brought together more than 90 different actors, including local governments,

BOX 1 GREAT GREEN WALL: BUILDING RESILIENCE

Ecosystem protection and rehabilitation is fundamental to building the resilience of food systems, particularly as climate change worsens. The Great Green Wall initiative is a major effort in the Sahel region intended to restore degraded landscapes across an 8,000 km strip of land between Senegal and Djibouti.[1] Initially, this ambitious pan-African program proposed constructing a 15-km-wide "wall of trees," but this goal was abandoned in favor of a more realistic mosaic of diverse landscape interventions, including natural regeneration, agroforestry, horticulture, livestock, apiculture, and water catchment infrastructure, in addition to reforestation.[2] Attention to the technical, social, and economic dimensions of this effort is essential to ensure success in improving environmental and socioeconomic outcomes.[3] However, a recent study showed that most of the restoration strategies designed in 12 participating countries to shape the Great Green Wall Initiative largely fell short in identifying potential benefits for different vulnerable or demographic groups, especially female-headed households and pastoralists, while potential risk for capture of the benefits by elite groups was not assessed.[4] On the financial side, it will require an estimated US$44 billion (under the base scenario) to fund all proposed land restoration activities, which would increase the economic value of Sahelian ecosystems over time – in terms of food, fodder, timber, and carbon sequestration – with an expected break-even point at most 10 years after implementation.[5]

UN agencies, NGOs, and donors, to align activities and promote collective outcomes.[33] The potential of the HDP approach to ensure greater coherence and impact in crisis responses has been highlighted by the recent establishment of the HDP Nexus Coalition hosted by the Global Network Against Food Crises.[34] However, implementation of HDP faces a number of constraints including limited understanding among actors in the three pillars of each others' roles, lack of joint analysis and scenario planning with in-country program teams, and the need for programmatic and financial flexibility in highly volatile contexts. It also requires negotiating trade-offs among the pillars – for example, engaging in conflict resolution may jeopardize basic humanitarian principles of nonpartisanship and thus impede access to vulnerable populations (see Chapter 3).[35]

REPURPOSING SUPPORT POLICIES to reduce the cost and increase the availability of nutritious foods will also be important for improving resilience and recovery from crises. The pressing question is how to finance a transition to better diets. Currently, official development assistance (ODA) for humanitarian purposes and crisis response is rising much faster than ODA for development purposes. As countries face more frequent or protracted crises,[36] African governments can expect increasing challenges in mobilizing new funding from both domestic and international sources to support their already underfunded development agendas.[37] The cost-effectiveness of investments will have to be improved, including by reallocating budgets and repurposing support policies. A recent scenario analysis[38] on repurposing existing public funding for food systems support showed potential for significant benefits in reducing the cost of nutritious diets, improving food security and nutrition, and reducing greenhouse gas emissions. However, trade-offs are also likely, including reductions in agricultural production and farm incomes. Thus, having complementary policies within and outside agrifood systems – such as social safety nets and affordable access to health services and education – as well as an environment for inclusive political participation will be needed to ensure that repurposing efforts lead to real improvements.

CONCLUSION

About 20 percent of Africa's population is food insecure and undernourished, more than double the population share in any other region of the world. Multiple crises in recent years – conflicts, natural disasters, disease, and economic shocks – have increased food insecurity across the continent. National and international actors, including governments, UN agencies, and NGOs, as well as affected local communities and households themselves are responding to the growing impact of crises in various ways, including through humanitarian assistance, early warning systems, migration, and resilience building. Crisis interventions that are responsive to gender are also critical to reducing disproportional impacts on women and girls. However, the costs associated with these responses are enormous and underfunded.

The HDP nexus approach offers a promising means to address the multifaceted nature of food crises more cost effectively in the short to medium term. For the longer term, however, repurposing current public support to food and agriculture will be critical to reduce the cost and increase the availability of nutritious foods. This multifaceted strategy to building crisis resilience over time would make healthy diets affordable and available for all of Africa's population, including the poor, women, children, and other vulnerable people, which aligns with African leaders' vision of accelerated transformation of food systems for shared prosperity and improved livelihoods. Systemwide enabling conditions for lasting resilience must include good governance mechanisms, adequate policies and regulations, high quality infrastructure, functioning community networks, and reliable safety nets.

MIDDLE EAST AND NORTH AFRICA

KIBROM ABAY, XINSHEN DIAO, DAVID LABORDE, AND MARIAM RAOUF

Kibrom Abay is country program leader for Egypt and a research fellow, Development Strategies and Governance Unit, International Food Policy Research Institute (IFPRI). **Xinshen Diao** is deputy division director, Foresight and Policy Modeling Unit, IFPRI. **David Laborde** is director, Agrifood Economics Division, Food and Agriculture Organization of the United Nations, and former senior research fellow, IFPRI. **Mariam Raouf** is a senior research associate, Development Strategies and Governance Unit, IFPRI.

While the global economy, and the economies of many countries in the Middle East and North Africa (MENA) region, has not yet recovered from the repercussions of the COVID-19 pandemic, 2022 brought new challenges triggered by the Russia-Ukraine war and associated trade shocks. The MENA region is particularly vulnerable to shocks to world food prices and trade because of its heavy dependence on food imports. It is also subject to political instability, fragility, and persistent conflict, all of which contribute to large refugee populations, many hosted by countries within the region, and to food insecurity more broadly. MENA is also among the world's regions most at risk from climate change and water scarcity.[1] The compound crises arising from conflict, trade shocks, and climate change currently threaten food and nutrition security in many MENA countries.

FOOD IMPORT DEPENDENCE AND RISING IMPORT COSTS

The MENA region relies heavily on food imports, especially cereal imports. For example, wheat represents 39 percent of caloric intake per person in Egypt, 20 percent in Sudan, and 46 percent in Yemen. Historically, much of this demand was met by imports from Russia and Ukraine.[2] In Egypt, the world's largest importer of wheat, imports account for about 62 percent of total wheat consumption, of which about 85 percent comes from Russia and Ukraine. Cereal import dependence is even higher in some other MENA countries, including Lebanon and Yemen.

At the onset of the current crisis, IFPRI researchers conducted an analysis of countries' vulnerability to the global increase in prices and the disruption of exports from Russia and Ukraine.[3] The country-level typology categorizes Lebanon, Sudan, and Yemen as extremely vulnerable to the crisis, and indicates Egypt is in the very high vulnerability category (Figure 1). For many countries in the MENA region, their direct exposure to the trade shock – as importers of Russian and Ukrainian cereals – and low existing stocks put their food security at risk. Existing stocks were already running low immediately before the crisis due to drought and crop failure.

Global food prices surged in early 2022 when Russia invaded Ukraine, disrupting Black Sea trade. Some exporting countries responded to these disruptions by introducing trade restrictions,[4] which put further pressure on global markets. Despite these challenges, many MENA countries have continued importing the usual volumes of food but at significantly higher prices (Figure 2), triggering a significant increase in import costs. For example, up to July 2022, MENA countries experienced a 50 percent increase in the cost of wheat imports. For some of these countries, the external crisis has been compounded by domestic production shortages, mainly due to weather conditions (Morocco and Iraq) and conflict (Syria), problems that have increased demand for imports just to meet basic consumption needs. Fortunately, most trade and

FIGURE 1 Overview of country-level relative vulnerability

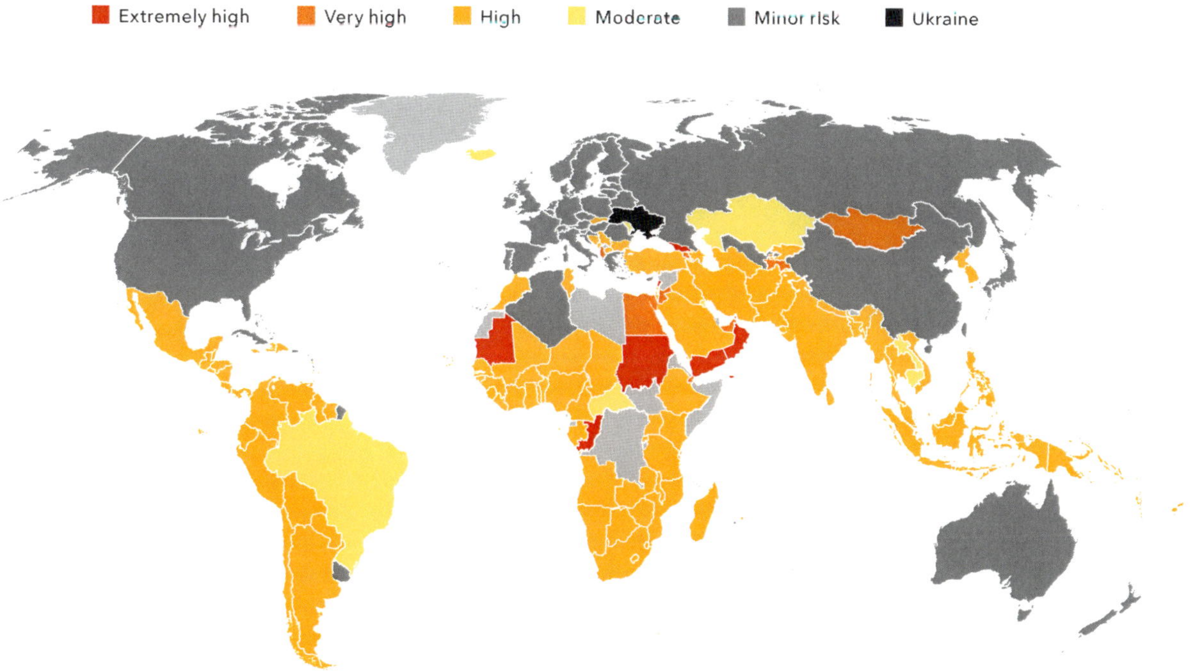

Source: Adapted from K.A. Abay, C. Breisinger, J. Glauber, S. Kurdi, D. Laborde, and K. Siddig, "The Russian-Ukraine War: Implications for Global and Regional Food Security and Potential Policy Reponses," *Global Food Security* 36 (2023): 100675.

Note: The indicators used for this assessment included: (1) existing dependency on the Black Sea region; (2) exposure to other suppliers that have implemented export restrictions; (3) current level of wheat stocks (to determine countries' buffer capacity); (4) consequences for countries' current accounts of price increases for various commodities (positive or negative effects depending on trade structure of countries); and (5) existing level of undernourishment, food price inflation, and expected impacts of the changes in world prices on domestic food bills and household food security.

financial sanctions continue to exempt food products and critical agricultural inputs like fertilizers. These exemptions may have forestalled a larger price increase for wheat.

WINDFALLS AND INCREASED ECONOMIC DIVERGENCE

In the face of global commodity shocks, the economies of MENA's oil-exporting countries have fared better than the region's oil-importing countries. The surge in oil and natural gas prices generated windfalls for MENA's oil exporters, although some of these countries, including Bahrain, Kuwait, Oman, Qatar, Saudi Arabia, and the United Arab Emirates, rely heavily on cereal and related food imports. MENA's oil-importing countries, such as Egypt, faced the double burden of high food and fuel prices. These price surges have raised import costs and reduced available government funds for oil importers,[5] triggering macroeconomic imbalances and major currency devaluations in Egypt, Lebanon, and Morocco. The devaluations in turn are causing significant inflationary pressure in domestic economies, which has fueled further price increases for a wide range of commodities and services in domestic markets. The surges in cereal prices have also significantly increased the cost of humanitarian assistance in fragile countries, such as Yemen and Sudan. For instance, rising wheat prices forced humanitarian organizations, including the World Food Programme, to reduce food-basket rations in both countries.

Within countries, the combination of rising fuel and food prices meant some sectors fared significantly better than others. As a result of the counteractive impact of the price increases for imports and exports, some countries' overall GDP and employment were affected less than initially expected. For example, while Egypt is a major wheat importer, it also exports natural gas and fertilizers. The windfall revenues from higher natural

FIGURE 2 *Wheat imports to MENA countries in 2022, compared to prior years*

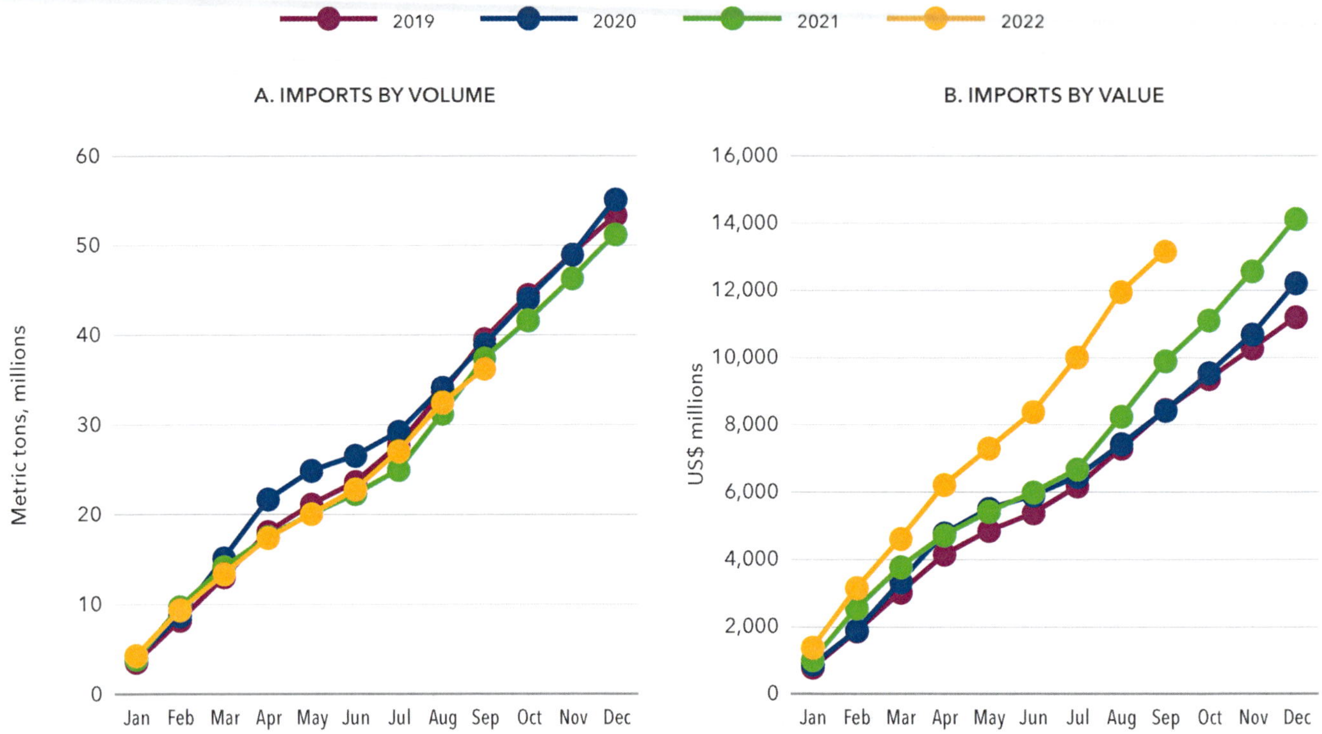

Source: Based on data from Trade Data Monitor (https://www.tradedatamonitor.com/).

gas prices have supported overall GDP, and firms and households with income associated with this sector are expected to benefit. However, Egypt's agrifood system has been harmed, particularly its off-farm agrifood system activities, which are energy-intensive (Figure 3).[6] Other fertilizer exporters faced more complex challenges. For example, Morocco is a large producer of phosphate, but relies on imports of intermediate inputs (either natural gas or ammonia) for fertilizer production. While high fertilizer prices could benefit Morocco, the war in Ukraine and the country's difficult relationships with neighboring countries, such as Algeria, complicated access to essential inputs in 2022.[7]

IMPACTS ON POVERTY AND INEQUALITY

Within countries, the crisis has had differential impacts across households, leading to an increase in inequality. Poorer households bear the greatest burden of current food price shocks because they spend a larger portion of their income on food and consume a disproportionate share of cereals and other cheap, energy-dense foods.[8] In Egypt, Sudan, and Yemen, for example, poorer households consume a significantly larger share of wheat-based calories per day than richer households.[9] Conflicts in Yemen and some other MENA countries further increase households' reliance on cereals and hence their vulnerability to food price shocks.[10]

The fuel price shock, in combination with the food price shock, is expected to further worsen inequalities. Windfall revenues from oil and natural gas exports are likely to accrue to governments, while most households – particularly poor or rural ones – are likely to be hit twice, by both rising prices and falling incomes. In Egypt, for example, overall national real household consumption is estimated to have fallen by a modest 0.9 percent (Figure 4), but rural and poor households have suffered a much larger decline in consumption than urban ones. Because Egypt produces most of the fertilizer it uses domestically and even exports a small amount, some urban households derive

FIGURE 3 Anticipated change in GDP and employment due to food, fuel, and fertilizer shocks in Egypt

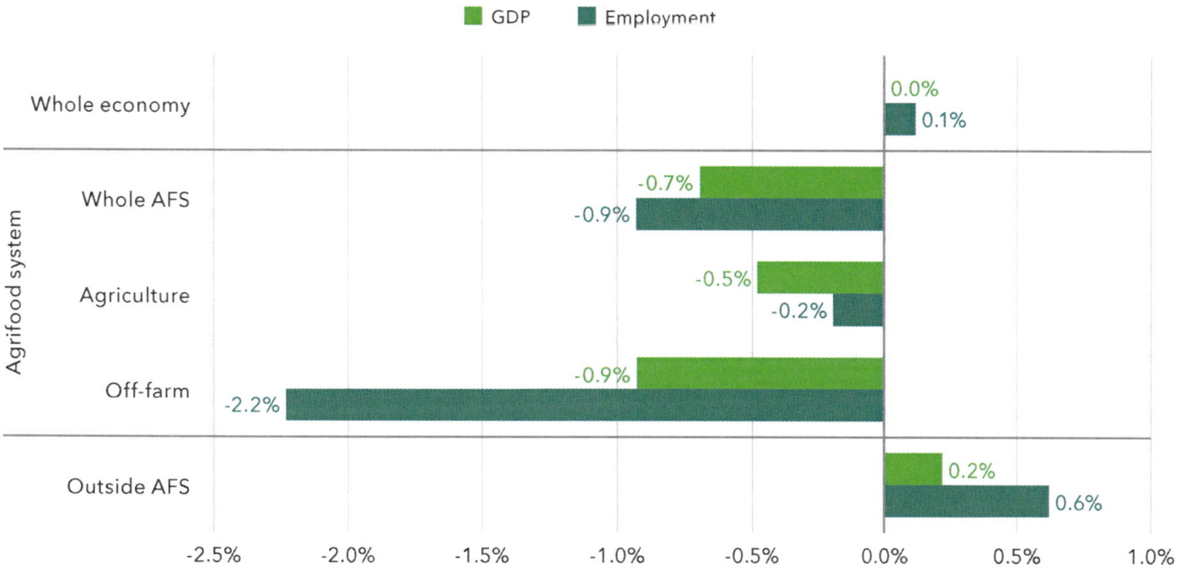

Source: Simulation results from IFPRI's Egypt RIAPA model, reported in K.A. Abay, F. Abdelradi, C. Breisinger, et al., "Egypt: Impacts of the Ukraine and Global Crises on Poverty and Food Security," Global Crisis Country Series Brief 18 (Washington, DC: IFPRI, 2022).

Note: Agrifood system (AFS) includes primary sector, food processing, and food-related services.

FIGURE 4 Anticipated change in real household consumption due to food, fuel, and fertilizer shocks in Egypt

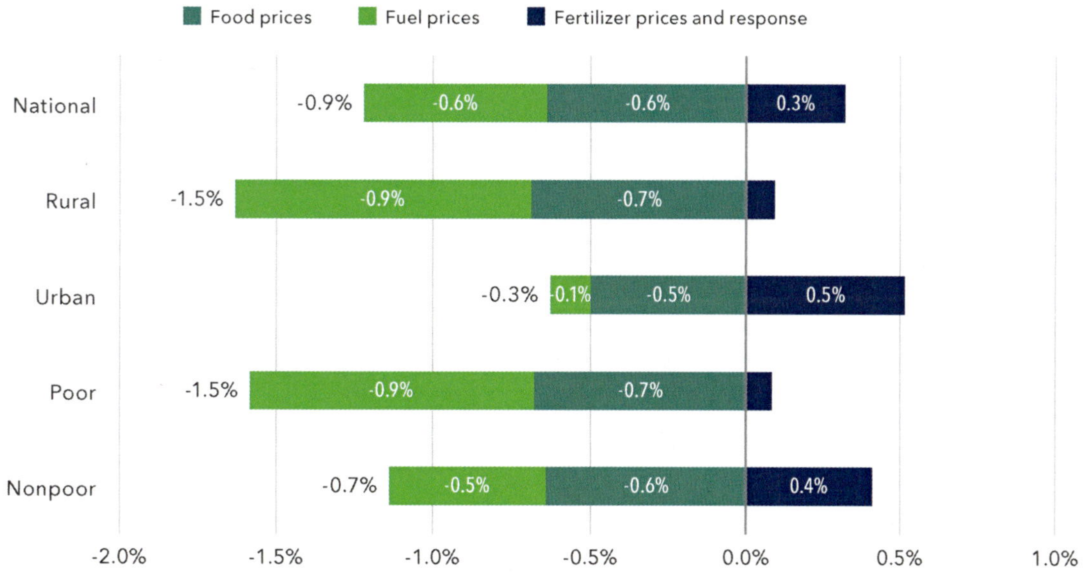

Source: Simulation results from IFPRI's Egypt RIAPA model, reported in K.A. Abay, F. Abdelradi, C. Breisinger, et al., "Egypt: Impacts of the Ukraine and Global Crises on Poverty and Food Security," Global Crisis Country Series Brief 18 (Washington, DC: IFPRI, 2022).

TABLE 1 Share of households affected by different types of shocks and food insecurity (2021–2022)

Country	Household type	Reported types of shocks affecting households						Prevalence of moderate or severe food insecurity (%)
		High food prices (%)	High fuel prices (%)	Sickness and accident (%)	Job loss (%)	Drought (%)	Other economic shock (%)	
Iraq	Male-headed	32	9	20	20	13	8	35
	Female-headed	35	8	34	21	8	7	49
Yemen	Male-headed	67	46	29	11	12	4	58
	Female-headed	62	40	46	6	6	5	74

Source: Based on FAO, Data In EMergencies (DIEM), accessed January 2023. https://data-in-emergencies.fao.org/pages/monitoring

income from fertilizer production and trade. As a result, the increase in fertilizer prices has had a positive impact for urban households as a group. Rural and poor households, however, have faced large impacts from all rising prices – for food, fuel, and fertilizer.

In several MENA countries, local conflict has compounded the impact of these global shocks. Countries affected by fragility, conflict, and violence saw the greatest increases in poverty caused by the COVID-19 pandemic.[11] Iraq and Yemen continue to grapple with the multiple shocks caused by conflict and high food and fuel prices, which all contribute to food insecurity. These underlying vulnerabilities are likely to affect households differently. For example, households headed by women in Iraq and Yemen are more likely to face idiosyncratic shocks such as sickness and accidents that reduce the income-generating potential of their households (Table 1). About one-third of households in Iraq and two-thirds in Yemen reported being affected by high food prices in the last two years, with those headed by women experiencing higher rates of food insecurity.

NATIONAL POLICY RESPONSES

The Russia-Ukraine war triggered important public policy responses, some of which have helped to contain inflationary pressures, though they have also contributed to fiscal pressures and costs. Several MENA countries introduced monetary and fiscal policies designed to cushion the adverse impact of the crisis on economies and households (Table 2).[12] Fiscal policies have included increased food and fuel subsidies, new price controls, incentives to boost domestic agricultural production, trade regulations, indirect tax exemptions, product-specific exchange rates, and the introduction or expansion of cash transfers and utility bill and financial support to vulnerable households. Some of these are adaptations of policies introduced in response to the COVID-19 pandemic. Others, including commodity subsidies, are new.[13] These measures have helped to limit price increases, but their medium-term impacts in terms of protecting households as well as the long-term fiscal implications for government debt remain to be evaluated.

PREPARING FOR COMPOUND CRISES

National policy responses to global food crises need to consider other regional vulnerabilities, including climate change, water scarcity, conflict, and rising debt vulnerability stemming from governments' increased fiscal spending. Recurring trade shocks and food crises are strong reminders that MENA countries need to reinforce their investments and efforts to increase the resilience of their food systems. In the very short term, MENA countries should consider diversifying their food imports and exports while continuing to invest in social protection systems to protect poor and vulnerable households from food price spikes. These social protection programs need to effectively target the most vulnerable groups, including women, who make up a large share of the poor.

TABLE 2 Public policy responses to mitigate the impact of trade shocks (introduced since February 2022)

	Product market interventions						Targeted social protection		
	Increased food and fuel subsidies	Instituted new price controls	Trade regulations	Indirect tax exemptions	Product-specific exchange rates	Increased regulated prices/ reduced subsidies	Cash transfers	Utility bill and financial support	Improved targeting
Algeria				●				●	
Bahrain				●				●	
Djibouti	●	●		●			●		●
Egypt	●	●	●		●		●		
Iran					●	●	●		
Iraq	●		●				●	●	●
Jordan	●	●	●	●		●	●		
Kuwait		●							
Lebanon					●	●	●	●	
Libya		●	●					●	
Morocco	●								
Oman		●						●	
Qatar									
Saudi Arabia		●					●		
Syria			●		●	●			●
Tunisia	●	●				●			
United Arab Emirates	●	●						●	
West Bank and Gaza	●	●		●					
Yemen			●						
Total: Out of 19	**8**	**10**	**6**	**5**	**4**	**5**	**7**	**7**	**3**

Source: Reproduced from F. Belhaj, R. Gatti, D. Lederman, et al. *New State of Mind: Greater Transparency and Accountability in the Middle East and North Africa–Middle East and North Africa Economic Update (October)* (Washington, DC: World Bank, 2022).

Note: These public policy responses, which are likely an incomplete list, were compiled by World Bank country economists. This list does not include monetary policy responses, such as increasing interest rates and devaluation, which have been deployed by some countries.

Targeting during the COVID-19 pandemic had mixed success in the MENA region, with targeting shown to be progressive (pro-poor) in some countries, including Egypt, but not in others, such as Morocco.[14] Rethinking consumer policies and adopting healthier and more sustainable diets (particularly reducing reliance on wheat) is also important. Indeed, while governments must prioritize protection for poor consumers in times of crisis, once prices have stabilized, they should focus on reforming food subsidies to improve diets and reduce vulnerability.

In the longer term, MENA countries will need to explore policy options for mitigating vulnerability to trade shocks that take account of domestic production capacities and constraints related to environmental sustainability and risk of weather shocks. Policies supporting the transition toward a greener future can offer a double win: less vulnerability to oil price shocks and a contribution to climate change mitigation. Given the region's strong potential for expanding wind and solar energy, it could generate additional revenues by diversifying exports.

Long-term agricultural policies in particular must take account of climate change and water scarcity. While some countries may have potential to expand arable land and production (such as

Sudan), such expansion is likely to be unsustainable in water- and land-scarce countries. For example, Egypt's principal focus should be on adapting its farming systems to address imminent water shortages and climate change threats and to increase resilience, rather than unsustainably expanding production.[15]

Windfall increases in state revenues in oil-exporting countries and the associated increase in government funds could serve all these objectives if fiscal surpluses are directed toward productive investments that diversify food imports and exports, thus strengthening the resilience of these economies. However, oil-importing countries, which continue to face significantly higher import bills and increasing debt vulnerability, need to devise sustainable means of addressing trade shocks and food crises. Those countries affected by prolonged conflict and violence should focus on restoration of livelihoods and protection of vulnerable households in the short term, while laying the groundwork for longer-term investment to support diversification and resilience of livelihoods.

CENTRAL ASIA

KAMILJON AKRAMOV

Kamiljon Akramov is a senior research fellow, Development Strategies and Governance Unit, International Food Policy Research Institute.

In Central Asia, the combined impact of the COVID-19 pandemic and the Russia-Ukraine war has ignited inflation and increased poverty. Although the region made good progress in reducing poverty and inequality over the past two decades, the pandemic stalled this progress and even reversed the welfare gains in some countries. Nearly half a million individuals in the region are estimated to have fallen into extreme poverty, due to decreased incomes, job losses, and work interruptions.[1] In Kyrgyzstan, for example, the poverty rate rose from about 20 percent in 2019 to more than 33 percent in 2021.[2] Subsequent external shocks to Central Asia's food systems, driven by the Russia-Ukraine war, have likely further worsened poverty and increased the vulnerability of households and individuals to food insecurity. Both these major shocks have constrained economic growth in the region. The Central Asian countries' strong trade and financial links with Russia and Ukraine, along with heavy reliance on remittances from their migrant workers in Russia, made them particularly vulnerable to the disruptions caused by the conflict, and the economic damage has been considerable.[3]

REMITTANCES, PRICES, AND FOOD SECURITY

Labor remittances play an important role in reducing poverty and inequality in Central Asia. In Kyrgyzstan and Tajikistan, remittances currently account for about 30 percent of national GDP, and more than 90 percent of these remittances come from Russia. Remittances also account for an essential share of income for many households in these countries. The impact of the war on labor migration and remittances has so far been mixed. Evidence suggests that migration interruptions following the war's onset were limited, and seasonal labor migration from the region between March and July 2022 increased slightly. However, the share of households with a member considering migration declined.[4] Data suggest the total flow of remittances to the region has been resilient and even increased significantly for Uzbekistan.[5] However, that growth cannot be explained by regular flows of remittances. Data from the National Bank of Kyrgyzstan suggest that while the total flow of labor remittances from Russia did not decline, the net inflow of labor remittances fell by nearly 14 percent in 2022 compared to 2021, with the declining trend more evident in the second half of 2022 (Figure 1). Unfortunately, we do not have data on the outflow of transfers from Uzbekistan.

Supply shortages and higher food and energy prices associated with the Russia-Ukraine war fueled double-digit inflation across the region. In Kazakhstan, annual inflation stood at 20.3 percent.[6] In Kyrgyzstan, overall inflation reached 14.7 percent in 2022, with food inflation at 15.8 percent and the consumer price index for wheat flour and products up 24.2 percent.[7] Since wheat and wheat products account for a significant share of caloric intake in the region, rising consumer prices could reduce household consumption, increase poverty further, and are likely to strike poorer households hardest. In Tajikistan, for example, wheat products account for about 45 percent of the average total caloric intake, and net wheat imports make up nearly 60 percent of the domestic supply (Figure 2). The retail price of wheat flour in Tajikistan rose

FIGURE 1 Monthly net inflow of remittances from Russia to Kyrgyzstan, 2021 and 2022

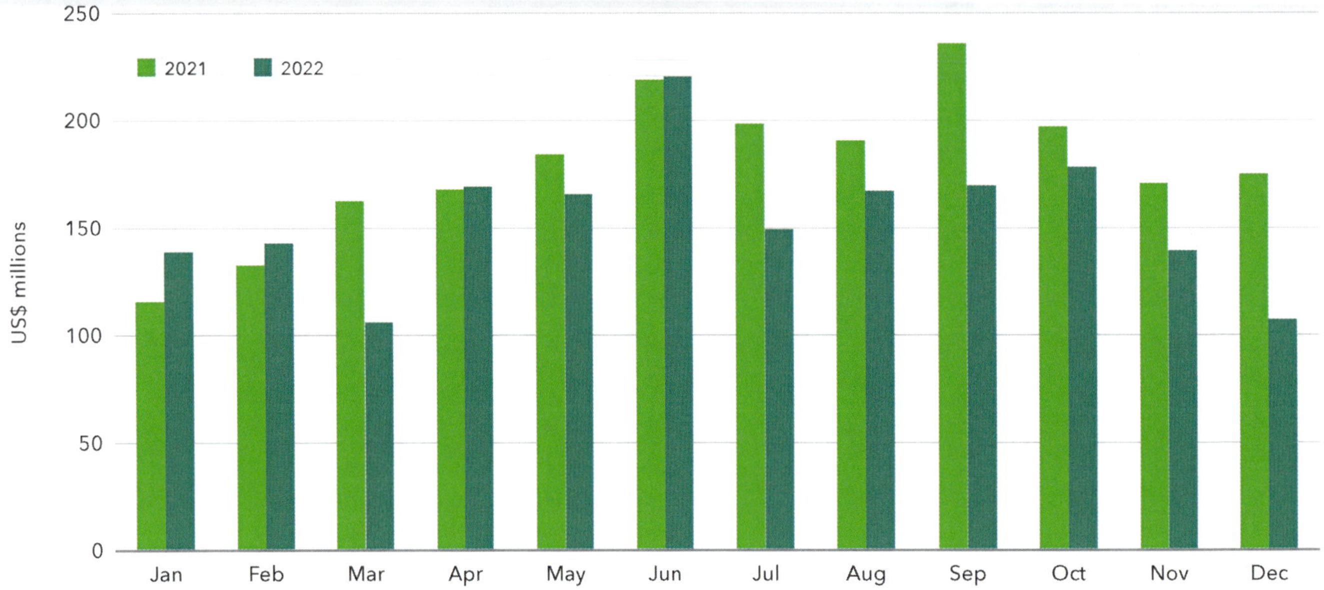

Source: Data from the National Bank of Kyrgyzstan (2023).

sharply in the first months of the war (February to May 2022), and despite a slight decline in the second half of the year, wheat flour prices remained more than 30 percent above the levels recorded at the end of 2021.[8]

LONG-TERM SOURCES OF FRAGILITY

In addition to the setbacks caused by the recent global shocks, food systems in Central Asia are at risk because of long-term sources of fragility, including gender inequality, climate change, and poor governance. Considerable gender gaps in labor force participation and earnings make women more vulnerable to external shocks and food insecurity during crises. When women enter their prime childbearing years, the gap increases as a result of their increased domestic and care burdens and the limited availability of public childcare services.[9] In addition, existing household, institutional, and societal gender inequalities add to the fragility of food systems in the region.[10]

Climate change poses a serious risk, given the large share of agriculture in GDP and employment in Central Asia. The region's agrifood sector and related livelihoods are exposed to increasingly

FIGURE 2 Share of wheat products in total caloric intake and net imports in domestic wheat supply, 2019/20

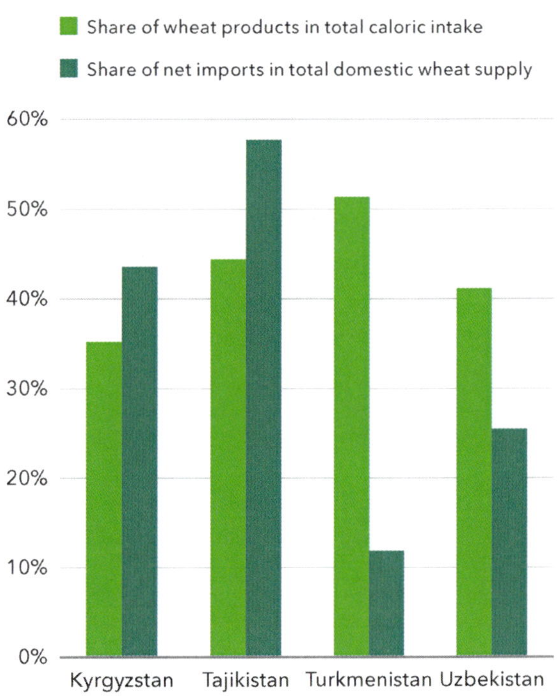

Source: Data from FAOSTAT (2022).

frequent extreme weather events, including temperature extremes, droughts, and floods, as well as greater variability in precipitation as global temperatures continue to rise.[11] For example, heatwaves in July 2021 and July 2022 and cold waves in January 2023 had major impacts on agricultural livelihoods and food systems in Uzbekistan.

The region's vulnerability to climate change is exacerbated by weak infrastructure, high levels of poverty, and poor governance. Despite some recent positive developments in governance, some evidence shows that weak political institutions, lack of accountability, poor regulatory quality and government effectiveness, and widespread corruption pose significant challenges to the stability of food systems in the region.[12]

LESSONS LEARNED

The recent COVID-19 pandemic exposed two significant weakness in Central Asia's food systems – a lack of diversity in markets and products and alarmingly weak governance. Central Asian countries score low on multiple dimensions of the World Bank's government effectiveness indicator, including the perceived quality of public services and the credibility of governments' commitment to their policies.[13] Poor governance and widespread corruption weaken Central Asian governments' capacity to collect revenue and spend public resources efficiently,[14] with detrimental impacts on their ability to respond adequately to external shocks and crises. Thus, public governance and anticorruption reforms should be a high priority in the region.

The lack of economic diversity is evident in the high concentration of imports from a few countries – for example, the Russian Federation, Kazakhstan, and China account for more than 50 percent of Uzbekistan's total imports – and dependence on remittances from a single country. In combination with rising inequality, this dependence on a few economic partners exacerbated the pandemic's negative impacts in the region, especially for poorer households. Increasing the number of trading partners and the diversity of supply chains, and economic diversification more generally, is essential to making the region's food systems more robust and resilient to external shocks (see Chapter 4).

Trade export restrictions, though not prolonged, caused considerable increases in food prices at the outset of the pandemic. Temporary bans and reductions in wheat exports imposed by the Russian Federation and Kazakhstan led to higher food prices in the wheat-importing countries – Kyrgyzstan, Tajikistan, and Uzbekistan. In response, some Central Asian governments shifted focus toward achieving a high degree of self-sufficiency in food, especially wheat. For example, in Tajikistan, policymakers have advocated for 80 percent self-sufficiency in grain, up from the current 45–50 percent they now produce. An IFPRI phone survey conducted in 2020 showed that many smallholders switched to growing wheat instead of high-value crops such as vegetables.[15] Central Asian countries are net exporters of vegetables, which allows them leeway to promote cereal production at the expense of these crops in order to reduce reliance on imported wheat. However, this shift could also cause food insecurity by reducing the dietary diversity that is accessible at affordable prices. Moreover, food self-sufficiency policies may require increased government intervention in agriculture, including price controls, subsidies, and regulation, which tend to create production and market inefficiencies and, as a result, may not achieve their desired outcome.

Social protection policies aim to protect vulnerable households and individuals from hardship caused by crises (see Chapter 5). In Central Asia, the social protection measures put in place during the COVID-19 pandemic were devoted to income protection, with a significant amount allocated to cash-for-work programs and unconditional cash transfers. However, they did not focus on job protection measures. Overall, these social protection policies were limited in scope. Moreover, weak governance and widespread corruption led to inefficient allocation and spending of limited public resources.[16] As a result, households resorted to negative food-based coping strategies, such as consuming less desirable, less expensive foods, as well as asset-depleting coping strategies.[17]

MONITORING AND RAPID RESPONSE

During the COVID-19 pandemic, regional governments and development partners worked together to monitor and respond to crises. These efforts were not sustained in the aftermath of the pandemic, leaving communities and households vulnerable to new shocks and failing to address the long-term impacts of the crisis on poverty, food security, and livelihoods. The World Food Programme has since established food security monitoring systems in Kyrgyzstan and Tajikistan, which conduct bimonthly household surveys to track trends in vulnerable communities. In Uzbekistan, the UN's Food and Agriculture Organization in partnership with Westminster International University in Tashkent recently launched a web-based monitoring tool designed to collect national food price data and facilitate its dissemination and analysis. However, these tools are limited in scope and focus on only a few aspects of food security. They are not designed to predict, monitor, or manage the long-term impacts of crises or vulnerability (see Chapters 2 and 3).

Policy responses to the COVID-19 pandemic differed widely across the region, largely reflecting governments' fiscal capacity. Kyrgyzstan and Tajikistan developed limited policy response measures, given their narrow fiscal space and limited public monitoring capacities.[18] Kazakhstan and Uzbekistan, with a larger fiscal base, responded swiftly to the crisis, implementing strict monitoring and confinement measures and designing large support packages. The multiple crisis response measures adopted by the Kazakh government included supporting the domestic private sector and employment; offering workers and families affected by the crisis short-term relief measures, such as cash payments to individuals who had lost their jobs or were on unpaid leave due to the quarantine; provision of food baskets and non-food essentials to vulnerable populations; and an increase in pension and social benefits.[19]

Uzbekistan's government developed and implemented a framework for local community-based monitoring and rapid crisis response. In addition to specific short-term measures intended to slow or prevent transmission of the virus and to ensure that health systems had the necessary capacity for response, this framework aimed to address the medium-term social and economic consequences of the pandemic. As elsewhere, the pandemic's impact was most severe for the poorest and most vulnerable. The government relied on the community-based targeting approach to reach the neediest sectors of the population as a part of the crisis response framework. But with a highly fragmented social protection system and limited government capacity, it was not able to deliver support to all vulnerable communities and households, and many needy households have received no assistance. In particular, the social protection coverage did not reach most unemployed and informal workers, leaving them more vulnerable to shocks. These poor households are forced to reduce consumption of nutritious foods, directly affecting their long-term nutrition, health, and productivity, with impacts that are difficult to reverse and perpetuate the cycle of poverty and vulnerability.

PREPARING FOR FUTURE SHOCKS

Several factors – including climate change, limited diversity of foreign trade, volatility of commodity prices, and dependence on remittances – make Central Asian countries especially vulnerable to external shocks and crises. The ad hoc approach that has been taken to managing such crises has failed to prevent serious increases in poverty, with long-term implications for development. Clearly, as crises become more frequent and even coincide, a more permanent, comprehensive framework for crisis readiness and response is needed. Such a framework will include a set of strategies, tools, and procedures put in place by the region's governments to prepare for emergencies and respond to them effectively by mitigating impacts and speeding recovery. Components may include risk assessment, early warning systems, a crisis management plan, communication, training and testing, and recovery and learning.

SOCIAL PROTECTION. Gender-sensitive social protection systems should be an integral part of any crisis response framework (see Chapter 6).

These programs can provide a safety net during short-term shocks as well as long-term changes in the labor market that affect incomes and jobs. An optimal safety net policy would protect the welfare of the poorest and most at-risk households and support sustainable growth without hindering the reallocation of labor to more productive sectors of the economy, which is essential for the development and transformation of food systems (see Chapter 5).[20]

Social protection systems should also cover Central Asia's labor migrants. Because these migrants work primarily in Russia, they are at risk when Russia's economy and labor market conditions deteriorate. With Russia subject to severe sanctions, labor migrants may lose employment because of declining demand, or fluctuations in the Russian ruble may make it difficult to exchange rubles for other currencies, such as the US dollar, thus decreasing the value of labor remittances. If these problems materialize, large numbers of migrants may return to their home countries, and they should be able to count on national social protection systems.

REGIONAL COOPERATION. Regional cooperation and foreign trade play a significant role both in making the relatively small and undiversified economies of Central Asia resilient to crises and in developing reliable national crisis response frameworks. However, the landlocked position of Central Asian countries is compounded by infrastructure bottlenecks, institutional and policy barriers, and poor trade facilitation, which need to be addressed. Trade and policy reforms, investments in physical and virtual connectivity, and cooperation in using shared natural resources, such as water resources, are essential to improve the region's readiness to respond to and prevent crises.

MONITORING AND ANALYSIS. Addressing increasingly frequent and disruptive crises will also require timely and well-tailored high-frequency data and analysis (see Chapter 2). At present, Central Asian countries collect little household and community-level data, and information and analysis is needed to identify the most vulnerable and affected populations and target social safety nets to them. The scarcity of gender-disaggregated data in particular makes it difficult to understand the differences and inequalities between men and women, address gender inequalities in crisis response, and ensure that policies and interventions are more effective in reducing gender disparities. Moreover, little in-country capacity exists for modern data analytics and assessment. Investment in gender-disaggregated, high-frequency data collection and in building analytical and applied research capacity is essential to better anticipate and prepare for future crises in Central Asia.

SOUTH ASIA

ANJANI KUMAR AND SHAHIDUR RASHID

Anjani Kumar is a senior research fellow and **Shahidur Rashid** is director, South Asia Regional Office, International Food Policy Research Institute, New Delhi.

The COVID-19 pandemic dealt a serious blow to the strong economic and social performance maintained by South Asia for two decades. Regional GDP shrank by almost 5 percent in 2020 (Table 1A). The agriculture sector, however, enjoyed modest growth across most of the region (Table 1B). As the regional economy struggled to recover from the pandemic, the Russia-Ukraine war and natural disasters, including devastating floods in Pakistan in 2022, led to further disruptions. Spikes in global food and energy prices and the tightening of global financial conditions, as countries tried to contain high inflation, led to contractions in South Asian trade and in the hospitality and manufacturing sectors. The deterioration in economic conditions that began with the pandemic led to a substantial increase in poverty, with 48–59 million people estimated to be newly poor in 2021, particularly in Afghanistan, Pakistan, and Sri Lanka.[1]

These shocks all contributed to higher food prices and disrupted food production and distribution. Food insecurity worsened in Afghanistan, Pakistan, and Sri Lanka during 2022 (Figure 1).[2] As pressures on food markets intensified, a number of food-exporting countries resorted to protectionist measures that subsequently contributed to higher international prices and market volatility. Afghanistan, Bangladesh, India, and Pakistan implemented export restrictions in 2022 on rice, wheat, and sugar, among other products. While international food prices have recently eased, they remain significantly above pre-pandemic averages, and continuing high prices for fertilizers and energy have made agricultural production less remunerative despite the increase in output prices.[3]

CURRENT ECONOMIC OUTLOOK

Economic prospects for the region are mixed, including for agriculture and food systems. In the second half of 2022, most of the region's domestic currencies depreciated by more than 10 percent against the US dollar (Bangladeshi taka, 18 percent; Pakistani rupee, 14 percent; Sri Lankan rupee, 45 percent), and consumer price inflation remains above national central bank targets.

India, which accounts for three-quarters of the region's output, showed robust growth of about 7 percent in 2022/23 despite recent shocks, and similar growth is expected in 2023/24.[4] Its agriculture sector also showed strong annual growth, at more than 3 percent. With this recovery, India is poised to become the fastest-growing economy among the world's largest emerging market and developing economies. Reasons for concern persist, however. Consumer inflation spiked to 7.8 percent in April 2022 and remained at 6.5 percent in January 2023, which led the Reserve Bank to tighten its monetary policy.[5]

Bangladesh was also hit by COVID-19 and the more recent shocks, although to a more limited extent than other South Asian countries. GDP growth is expected to slow from 7.2 percent to 5.2 percent in 2022/23 due to falling exports, a growing trade deficit, continued high inflation, reduced remittances, energy scarcity and higher prices, and tighter monetary policy.[6]

Pakistan – an already vulnerable economy with a debt equal to 97 percent of its GDP,[7] soaring inflation, and acute shortage of foreign exchange reserves – faces continued policy and political uncertainty as well as damage from natural

TABLE 1 Annual GDP and agricultural GDP growth in South Asia

	2019		2020		2021		2022
Country	GDP growth	Ag GDP growth	GDP growth	Ag GDP growth	GDP growth	Ag GDP growth	GDP growth
Afghanistan	3.9	17.5	-2.4	5.9	-20.731	-2.8	NA
Bangladesh	7.9	3.3	3.4	3.4	6.9	3.2	7.2
Bhutan	4.4	1.3	-2.3	4.6	-3.3	2.1	4.0
India	3.7	5.5	-6.6	3.3	8.7	3.0	6.8
Maldives	6.9	-7.5	-33.5	7.1	37	-0.6	8.7
Nepal	6.7	5.2	-2.4	2.4	4.2	2.8	4.2
Pakistan	3.1	0.9	-0.9	3.9	5.7	3.5	6.0
Sri Lanka	-0.2	0.5	-3.5	-1.4	3.3	2.5	-8.7
South Asia	**3.8**	**4.8**	**-4.8**	**3.4**	**8.1**	**3.0**	**6.4**

Source: GDP growth data are from the International Monetary Fund's World Economic Outlook (2023); agricultural GDP growth data are from the World Bank's World Development Indicators (2023).

Note: NA indicates data not available.

disasters that pushed an estimated 5.8 to 9.0 million people into poverty in 2022.[8]

Sri Lanka and Afghanistan are also still facing crisis. Sri Lanka's output fell by an estimated 9.2 percent in 2022 and is expected to decline another 4.2 percent in 2023 as a result of ongoing foreign exchange shortages, high inflation, increased interest rates, and policy measures implemented to restore macroeconomic stability. This economic crisis increased poverty and reversed income gains made over the past decade. In Afghanistan, the sudden suspension of international aid in August 2021, along with reduced foreign investment, shrank the country's output by about one-third, leading to a large increase in poverty. The situation there remains precarious, and severe food shortages are likely.

On the other hand, Nepal has enjoyed a strong recovery in domestic demand, which may raise GDP growth to 5.8 percent in 2022/23, and the country is expected to maintain robust growth for the foreseeable future. In addition, the Maldives and Bhutan are benefiting from the post-pandemic recovery of tourism. The Maldives is likely to remain the fastest-growing small economy in the region due to infrastructure investments and the rebound in tourism. Bhutan's economy is projected to grow by 4.1 percent in 2022/23, as a result of opening its border with India in September 2022.

FOOD SECURITY AND POVERTY

South Asia is far off track to achieve Sustainable Development Goal 2 (SDG2), Zero Hunger, by 2030, and progress in tackling the problem has stalled. Numbers of undernourished people and those facing severe food insecurity are up substantially over the past five years (Figure 1). Child stunting and wasting remain more prevalent in South Asia than in other world regions. The deterioration in food security is largely due to the pandemic-induced economic disruptions, poor macroeconomic management, armed conflicts, and climate change. Progress toward SDG1, No Poverty, has also been set back, as hard-won gains have been lost and the pandemic pushed an additional 62–71 million people into poverty in South Asia.

The recovery and development of food systems in South Asia face multiple challenges. Although spillover effects from the Russia-Ukraine war have not been large, South Asia has been affected by the global rise in food, fuel, and fertilizer prices. Food prices have risen sharply, contributing to food insecurity. In September 2022, the year-on-year consumer inflation rate for food was 66 percent in Sri Lanka, 36 percent in Pakistan, and about 8 percent in India, Bangladesh, and Nepal. The inflation in Pakistan and Sri Lanka is attributed mainly to macroeconomic instability and mismanagement, especially the sharp devaluation of their currencies, and the fertilizer ban in Sri Lanka.

NATURAL DISASTERS

Climate change is another significant threat. Diverse geophysical settings and climatic conditions make the region vulnerable to various environmental shocks.[9] Natural calamities, many related to climate, have become increasingly frequent over the past two decades (Figure 2), with a corresponding increase in the numbers of people affected in many countries. Several extreme weather events occurred in 2022, compounding the other shocks to the region. Record-breaking heatwaves in Afghanistan, Bangladesh, India, southern Nepal, and Pakistan posed serious threats to life, livelihoods, and economies.[10] In Pakistan, severe droughts followed by devastating floods inflicted major damage on agricultural production. These back-to-back catastrophes affected approximately 33 million people. Economic losses are estimated at US$15 billion, and the country's GDP declined by about 5 percent.[11] Pakistan's federal and provincial authorities are now working with local, national, and international partners to manage massive relief efforts across the country.

Afghanistan suffered two major earthquakes in 2022 that affected about 9,000 people. In response, the government allocated $11.3 million for disbursement to the affected population, including the injured and households that lost family members. Adding to this misery, Afghanistan suffered a drought that affected 80 percent of the country, and production of wheat declined as a result of the 2022 La Niña occurrence, which stressed the country's water resources, adding to food insecurity.

India is prone to many major natural hazards, and in 2022 recorded a broad range of extreme weather events that caused more than 3,000 human deaths and 60,000 animal deaths, and damaged 2 million hectares of crops. Erratic monsoon rains led to increased food price volatility, threatened households' inflation expectations, and complicated monetary policy management.[12]

Bangladesh, too, is extremely vulnerable to natural disasters. In 2022, its northeastern region suffered a devastating flashflood that affected about 7.2 million people. Timely and appropriate crisis response is increasingly important amid continuing climate change, as yields for rice, vegetables, and wheat are expected to decline by 5 to 6 percent by 2050.[13]

Other countries in the region are likewise at risk. Nepal is at high risk of earthquakes. Sri Lanka experiences a high incidence of disasters relative to its small size and concentrated economic activities, with average annual disaster-related losses of $50 million, affecting some 500,000 people.

South Asian countries are learning from past disasters to improve responses. For instance, in the aftermath of the 2004 tsunami in India, the 2005 Enactment of Disaster Management Act sought to minimize future losses by integrating disaster management measures at all levels of governance, including national, state, and district-level authorities. In addition, an Early Warning System for Tsunamis in the Indian Ocean was established by the Indian government in 2007 to issue advance warnings in coastal areas, which could reduce impacts of future disasters.[14] Similarly, timely policy initiatives taken by the Government of Bangladesh after the devastating floods in 1998 – including enabling private sector participation in grain markets and enhanced public investments in agriculture – have helped respond to subsequent shocks.[15] The relatively low death toll and low incidence of waterborne diseases after flashfloods in Bangladesh in 2004 reflect the efficacy of the country's disaster preparedness and response capabilities, and people's ability to manage and recover from disasters. These efforts have borne fruit and therefore, despite the frequency of natural disasters in South Asian countries, governments in these countries have been able to respond to recent shocks more effectively than in the past.

POLITICAL INSTABILITY AND CONFLICTS

Political instability and violence also threaten food security in the region. Since their independence, many South Asian countries have experienced political instability caused by civil wars and ethnic and sectarian conflicts. As a result, a sizable number of people have been displaced.

FIGURE 1 Undernourishment and severe food insecurity in South Asia

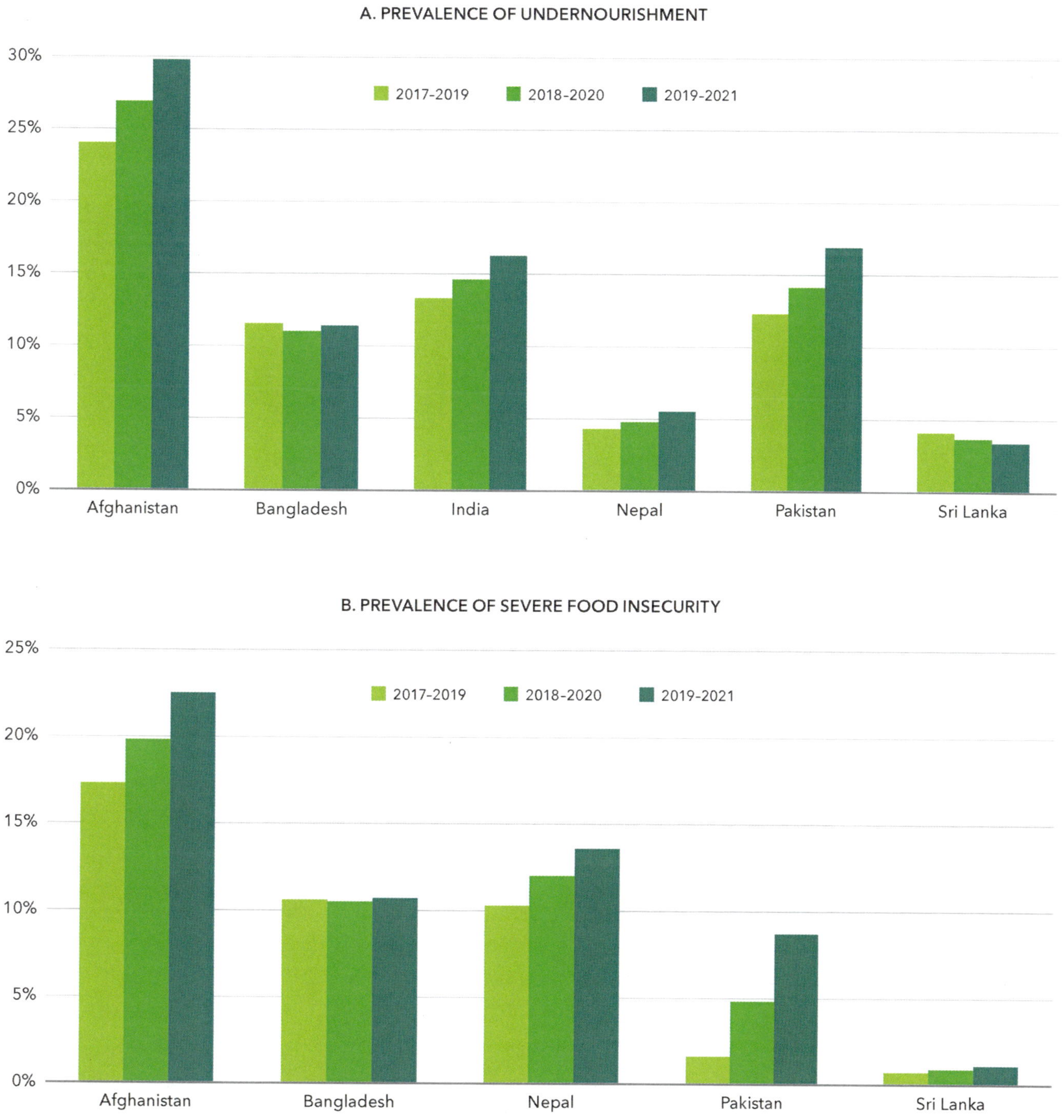

Source: Data from FAOSTAT, accessed Jan. 2023. https://www.fao.org/faostat

Note: The prevalence of severe food insecurity is an estimate of the proportion of the population that resides in severely food-insecure households. The assessment is conducted by using data collected with the Food Insecurity Experience Scale (see Chapter 2) or a compatible experience-based food security measurement questionnaire. A household is classified as severely food insecure if at least one adult has reported several of the most severe experiences described in the FIES questions, such as being forced to reduce the quantity of food, skipping meals, and going hungry due to lack of resources. Measures for severe food insecurity are not reported for India.

FIGURE 2 Natural disasters in South Asia, 1980-2022

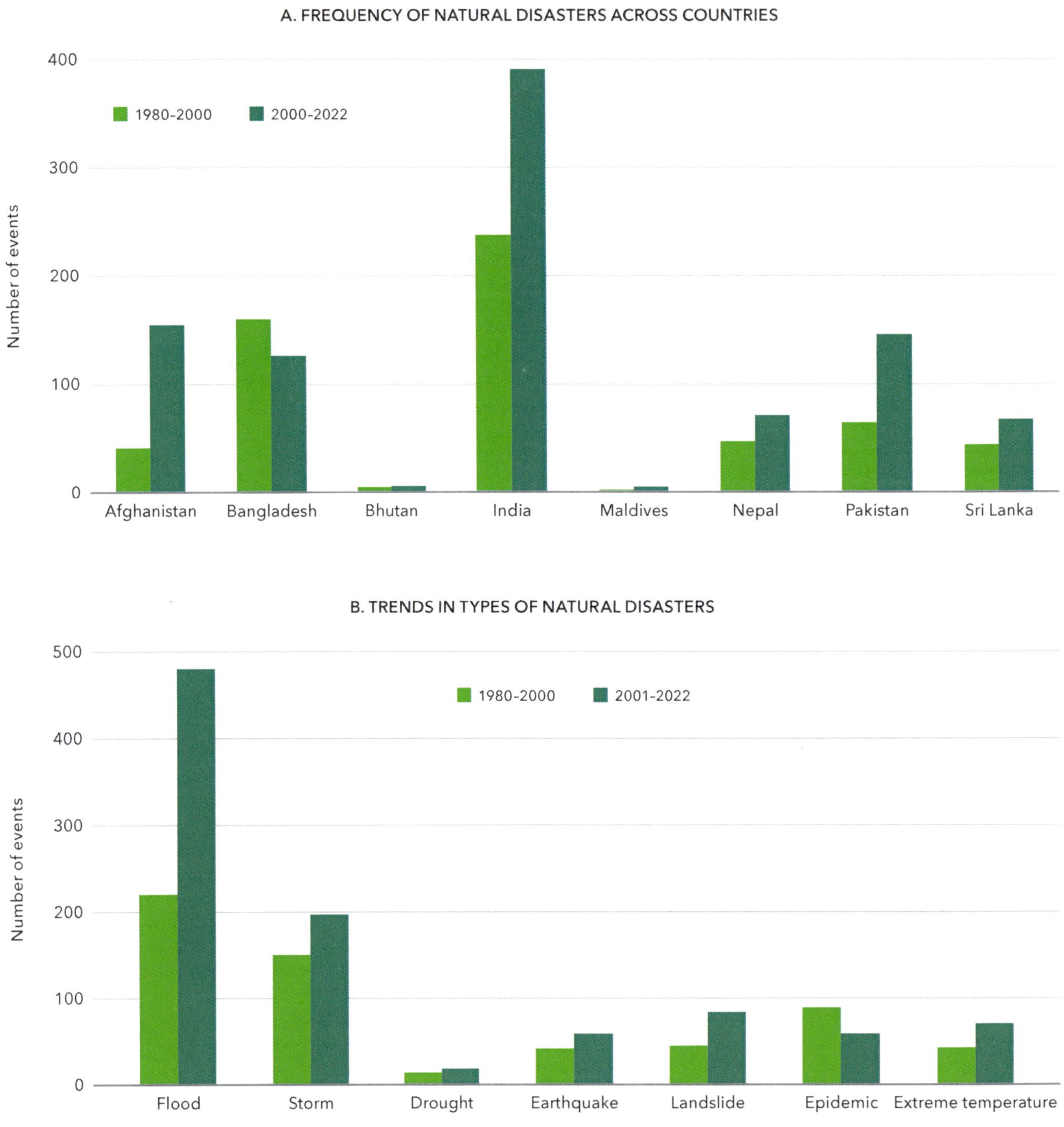

Source: Data from EM-DAT, accessed January 2023. https://www.emdat.be/

Afghanistan has been affected by the Taliban insurgency and other forms of violence for decades. The political crisis after August 2021 led to a sharp economic contraction (Table 1), rising food insecurity, and an increase in poverty.[16] The Rohingya crisis continues to pose serious challenges to Bangladesh's government, in collaboration with various international agencies, in coping with the enormous influx of refugees that has made Cox's Bazar the world's largest refugee camp. Additionally, violence resulting from the government crackdown on the opposition party in December 2022, ahead of a major rally, further added to internal disruptions in Bangladesh.

Nepal has a long history of political unrest, but a new constitution, drafted in 2015, established a federal structure in the country, fostering renewed hope for greater political stability, social cohesion, good governance, and sustainable development.

Sri Lanka faces a volatile political situation, exacerbated by the country's unsustainable debt and a severe balance-of-payments crisis. With declining economic growth and increasing poverty, political and economic instability are expected to continue.

RECOMMENDATIONS FOR FOOD CRISIS RESPONSES

A combination of short- and long-term measures are required to tackle food system crises in South Asia. These include:

- Identification of vulnerable households and groups (women, children, the elderly, and disabled persons) to provide them adequate support. The current food shock could be used as an opportunity to strengthen social safety nets, and targeted food consumption subsidies could be explored to gradually replace broader food consumption subsidies.

- Promotion of intraregional trade, including removal of recently adopted protectionist policies, given that trade within South Asia is quite limited compared with other regional blocs.

- Increasing production by improving smallholders' access to modern technologies and inputs – especially for women and other disadvantaged groups – including facilitating access to fertilizers, promoting crop diversification, and boosting innovative technologies and approaches.

- Stepped-up investment in customized climate-resilient agriculture for longer-run sustainability, which can be supported by repurposing existing agricultural support.

- Long-term systematic preparedness to mitigate disruptions in food systems, including strategic and resilient food security programs. Other South Asian countries could gain valuable insights from India's National Disaster Management Authority and One Nation One Ration Card and from Bangladesh's National Action Plan for Food Security and the effective implementation of its Food Friendly Programme.

Beyond these measures, South Asian countries should align with international development agencies for funding support to build resilience in the agrifood system. They should also develop a long-term strategic framework to address the macroeconomic mismanagement in Pakistan and Sri Lanka, conflict in Afghanistan, and the refugee crisis in Bangladesh, and establish a continual effort to improve governance.

EAST AND SOUTHEAST ASIA

KEVIN CHEN, YUNYI ZHOU, AND RUI MAO

Kevin Chen is a senior research fellow, Development Strategies and Governance Unit, International Food Policy Research Institute (IFPRI), Beijing, and a Qiushi chair professor, Zhejiang University, Hangzhou, China. **Yunyi Zhou** is a graduate research assistant, Zhejiang University-IFPRI Center for International Development Studies, China Academy for Rural Development, Zhejiang University. **Rui Mao** is a professor and deputy director, Zhejiang University-IFPRI Center for International Development Studies, China Academy for Rural Development, Zhejiang University.

Amid a global recession and the Russia-Ukraine war in 2022, East and Southeast Asian countries have experienced setbacks on their path toward meeting the Sustainable Development Goals (SDGs). Nonetheless, the region's trade and financial positions have been relatively unaffected as yet by the war, compared with much of the world. For 2022, economic growth is expected to average 3.8 percent in East Asia and 5.0 percent in Southeast Asia.[1] Threats remain, however, as climate-related disasters, the pandemic, economic slowdown and protectionism, and their nested repercussions are unlikely to ease in the short term and could further disrupt the region's food systems in 2023.[2] Yet intraregional integration has continued to deepen, which can be expected to bolster the region's resilience to crises, and the UN's 2030 Agenda calls on countries in the region to collaborate in creating a globally competitive, integrated, resilient, and inclusive food system that will be better positioned to weather future crises.[3]

DISRUPTED PROGRESS TOWARD ENDING POVERTY AND HUNGER

Despite some economic recovery in 2022, progress toward achieving SDGs 1 and 2 – No Poverty and Zero Hunger – has been disrupted. Across the region, there are huge disparities in food and nutrition status, which have increased in recent years. In East Asia, a rise in severe food insecurity in 2020 was reversed in 2021; in Southeast Asia, both the absolute number and the percentage of people facing severe food insecurity increased in 2020 and 2021 (Figure 1). These trends are reflected in the Global Hunger Index for 2021, where East Asia scored well but Southeast Asian economies overall fared worse. In 2022, Myanmar, Cambodia, and the Philippines had the highest rates of insufficient food consumption among member states of the Association of Southeast Asian Nations (ASEAN).[4]

On a more positive note, the incidence of poverty and the number of poor in East and Southeast Asia (except China) in 2022 are projected to return to the levels that were forecast pre-COVID-19, although progress has been fragile. Low-income households (and especially women, children, and the elderly) are vulnerable to the food and energy price increases caused by global supply chain disruptions and the recent war.[5] The cost of a healthy diet rose in both subregions between 2019 and 2020, and inflation may have put healthy diets even further out of reach in 2021 and 2022.[6] Moreover, no country in either subregion is on track to meet the targets for curbing adult obesity or anemia in women of reproductive age.[7]

KEY VULNERABILITIES AND RESPONSES

COVID-19 PANDEMIC

For more than three years, the repeated COVID-19 shocks have affected demand, supply, and trade in the region's food systems.[8] Widespread vaccination in the region and major economies elsewhere

FIGURE 1 Headcount and prevalence of severe food insecurity

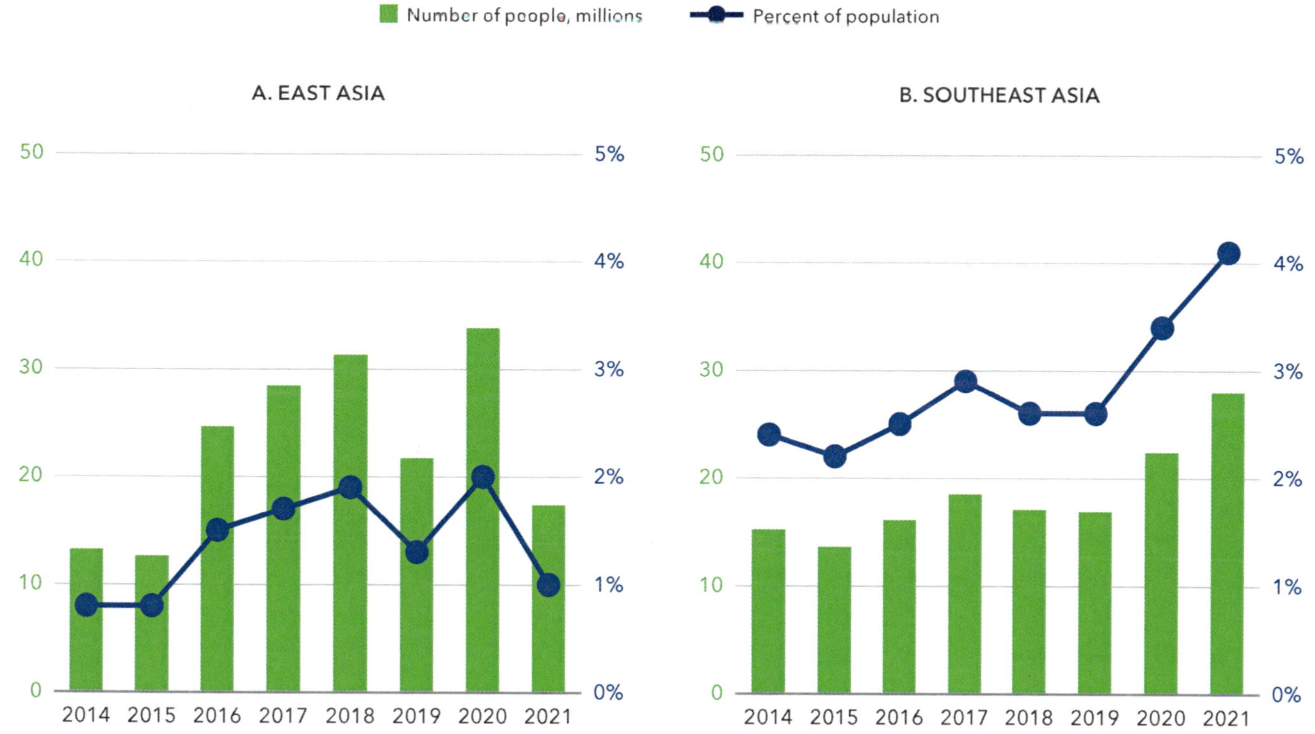

Source: Adapted from World Bank, *East Asia and Pacific Economic Update, October 2022: Reforms for Recovery* (Washington, DC: 2022).

has allowed East and Southeast Asian governments to gradually shift their policy focus from managing pandemic disruptions to supporting post-pandemic recovery.[9] As pandemic-related restrictions were gradually lifted in 2022, many Southeast Asian economies began to revive. Recovering international tourist arrivals are expected to help countries such as Thailand and the Philippines recapture lost revenue along with jobs in food services and many other sectors.[10] China began loosening its pandemic policies in late 2022 and is increasing its pro-growth stance. Despite recent challenges in the public healthcare system as COVID-19 cases rose, its economy is expected to return to buoyant growth in 2023 as a result of reopening and possible policy stimulus, with positive impacts on global value chains.

Among ASEAN member states, disparities in income and access to public services between rural and urban areas and between men and women worsened during the pandemic, suggesting that low-income and marginalized households (such as informal employees, migrants, and rural populations) will be more susceptible to long-term setbacks and inequalities during recovery (for example, lower savings and scant access to credit and jobs). Addressing these disparities will require a more inclusive financial system in the wake of the pandemic.[11] In addition, many East and Southeast Asian governments increased unsustainable measures, such as environmental deregulation, in response to the pandemic disruptions.[12] Coordinated action to reinforce food system resilience in the face of climate change and biodiversity loss is essential for sustainable post-pandemic recovery.

To cushion the socioeconomic impact of COVID-19, most nations provided "rescue packages" (such as in-kind food distribution, cash transfer programs, and expanded social protection) along with targeted measures to support domestic food production and consumption. For example, in Thailand, where two-thirds of laborers work in the informal sector, the government responded to the outbreak in 2020 with fiscal packages designed to support small and medium enterprises (SMEs), farmers, and informal employees outside the social security system.[13] Across the region, measures that aimed to sustain the food supply and protect producers included increased agricultural input subsidies and distribution, price support through procurement and regulation, new programs stimulating local food production and short value chains, and broad-based rural development policies.[14] For example, Malaysia allocated about US$225 million through the Bank Negara Malaysia Agrofood Financing Scheme to improve agricultural productivity and encourage local food production, with approximately $4.5 million earmarked to train more small farmers in using digital technologies.[15]

CLIMATE CHANGE

Myanmar, the Philippines, and Thailand remained among the countries deemed most at risk by the Global Climate Risk Index in 2021.[16] China experienced a mix of record-breaking heatwaves, severe drought, and heavy rainfall in 2022, causing a decline in its annual grain yields (relative to projections based on prior-year trends), although the country has sufficient domestic reserves to buffer the impact on food supplies.[17] But as global warming worsens, the adverse impacts of extreme weather events on food security will rise, far outweighing the potential increase in output of some crops due to warmer temperatures.[18] On the consumption side, the demand for food (especially animal-source foods, maize, and soybeans) is increasing as a result of population growth, rapid urbanization, and rising household incomes. The region's reliance on conventional approaches to agricultural productivity growth (for example, its synthetic nitrogen fertilizer application rates are among the highest in the world) will make shifting to low-carbon development of the sector challenging.[19]

Many countries in the region have not yet set agriculture-specific targets for greenhouse gas emissions, though Viet Nam has committed to reducing its emissions by 20 percent every 10 years and is building its capacity for measuring, reporting, and verifying farm-level emissions.[20] To improve agricultural productivity within the bounds of sustainability, many countries have supported climate-smart agriculture technologies and practices, including capacity building, climate-resilient crops, planting calendar adjustments, and more efficient machinery. For example, the China Weather Index Insurance Project offers digital insurance that has shown promise for stabilizing the income of small-scale farmers facing losses from natural disasters.[21] A cross-country review of experiences with scaling out location-specific climate-smart agriculture models in ASEAN recommends starting with knowledge sharing, then mainstreaming the tested interventions into government policies, and finally sustaining efficiency with proper market strategies.[22]

ECONOMIC UNCERTAINTIES

The Russia-Ukraine war has had only modest direct impacts on food systems in East and Southeast Asia compared with other regions. Deepened intraregional trade and value chain participation, stable rice production and inventories in the region, and relatively limited wheat consumption in the Southeast subregion are important factors accounting for this resilience. However, the widening impact of the war and global inflation, along with climate change, have been driving food and livelihood crises in the region. No country has escaped recent food inflation, with food insecurity of most concern in countries highly dependent on food and agricultural input imports. The rapid rise in fertilizer prices (especially nitrogen and phosphates) – induced by the war in Ukraine, reduced fertilizer production in the European Union, and a contraction in fertilizer production and exports from China – has raised concerns about the region's food security and potential systemic economic crises.[23] On the other hand, Viet Nam and Thailand are projected to increase rice exports in 2023 in response to strong demand.[24]

Many governments in the region increased support to agricultural production and even restricted exports in an effort to tame domestic pressures on food and fuel prices during the pandemic. However, those short-term measures (mostly public policy support through price controls and trade barriers) distorted markets and disrupted the trend toward green production and dietary diversification (Figure 2). Further price inflation affecting

FIGURE 2 Public policy support to agriculture, 2010–2020 average

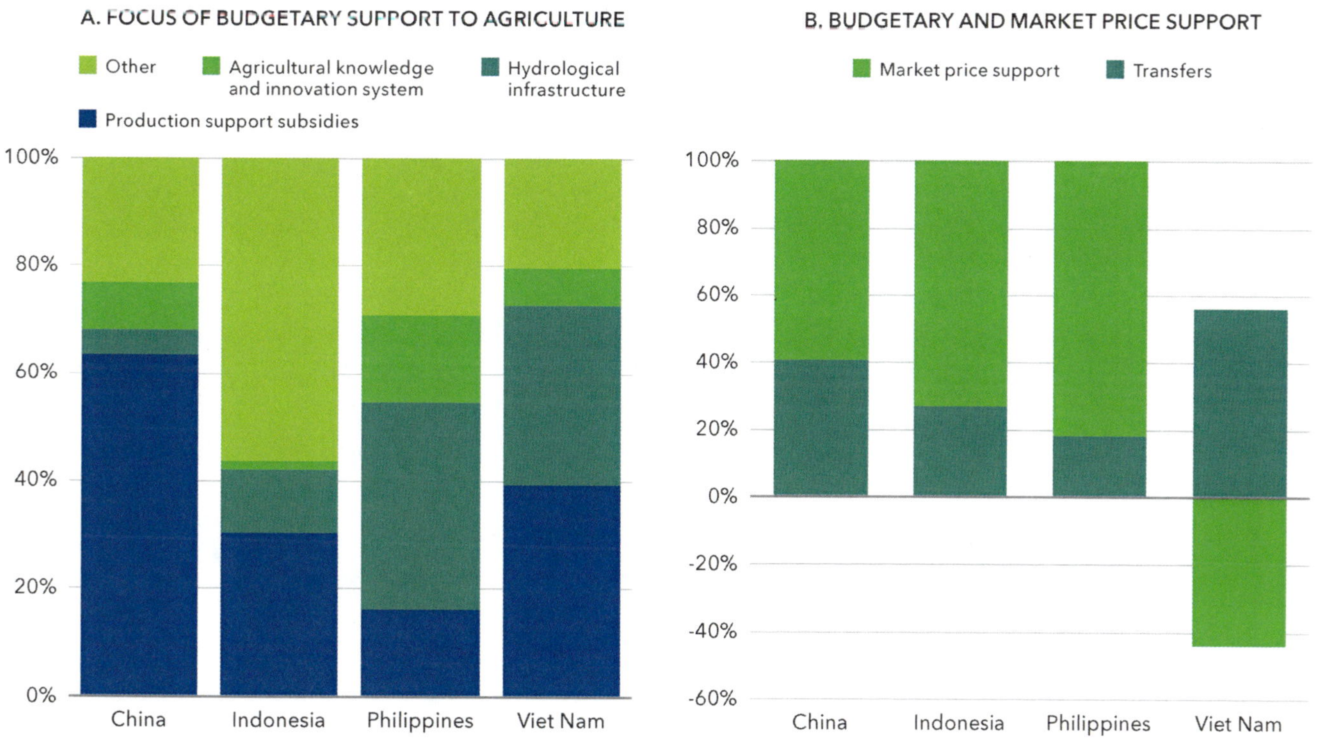

Source: Reproduced from World Bank, *East Asia and Pacific Economic Update, October 2022: Reforms for Recovery* (Washington, DC: 2022).

agrifood commodities is likely to increase the budgetary costs of agricultural input subsidies and food price controls, limiting governments' scope for further policy support in agriculture. Based on evidence from Thailand, cash transfers to vulnerable groups are recommended as a more cost-efficient alternative to price interventions for supporting food security.[25] With differing governmental capacity (fiscal positions) to sustain fiscal buffers, the agrifood sector in the Philippines, Thailand, and Malaysia may be most at risk from reduced agricultural input subsidies and food price support.[26]

INTEGRATION FOR THE FUTURE

Considering growing fiscal deficits, food inflation, and debt, substantial work is needed to put the region's agrifood systems on track toward resilience and sustainability, especially in a gloomy global economic environment.[27] Several major integration frameworks can help build resilience for the region's future. The 2020 ASEAN Comprehensive Recovery Framework highlights the development of climate-smart agriculture and the need to boost agro-rural productivity. The 2021 Global Call to Action for a Human-Centered Recovery, from the International Labour Organization, provides a framework for proposed actions within ASEAN member states. The ASEAN-China Joint Statement on Enhancing Green and Sustainable Development Cooperation, also announced in 2021, is expected to expand actions to move food systems toward the SDGs. In addition, the Regional Comprehensive Economic Partnership (RCEP), which came into force at the beginning of 2022, could galvanize regional integration and enable ASEAN member states and their East Asian partners to better manage a complex array of food system crises and build resilience for the future through a multilateral trading system.

LATIN AMERICA AND THE CARIBBEAN

EUGENIO DÍAZ-BONILLA AND VALERIA PIÑEIRO

Eugenio Díaz-Bonilla is special advisor, Inter-American Institute for Cooperation on Agriculture, and a visiting senior research fellow, International Food Policy Research Institute (IFPRI). **Valeria Piñeiro** is acting head of the Latin America and Caribbean program and a senior research coordinator, Markets, Trade, and Institutions Unit, IFPRI.

The countries of Latin America and the Caribbean (LAC), like most of the world, have been affected by multiple economic, health, and geopolitical shocks in recent years, all adding to the damage from more frequent extreme weather events. This section reviews major impacts from these crises, which have varied across the LAC region, reflecting the wide variation in national economies, and offers recommendations for reducing the impact of future shocks.

Historically, the region's economies have been substantially affected by global commodity cycles, which drive high economic volatility; in fact, the variability in the per capita growth rate in LAC is about double that of East and South Asia (although lower than in sub-Saharan Africa and the Middle East and North Africa). Economic volatility in turn has exacerbated the negative economic and social consequences associated with LAC's large income inequalities (LAC and sub-Saharan Africa are the two regions with the highest levels of inequality). On the positive side, democratic governance is more common in LAC than in other developing regions. However, the combination of economic volatility and inequality has affected the quality of democracy and the functioning of governments.[1]

Figure 1 illustrates the critical role of commodity cycles for Latin American economies, beginning with the high commodity prices and economic growth during the 1970s, when income per capita grew at about 3.7 percent per year. This was followed by the drop in commodity prices and the "lost decade" of the 1980s, when average GDP growth was negative (−0.6 percent from 1981 to 1990). The most recent commodity cycle began in the first half of the 2000s, with commodity prices peaking around 2011 and then declining until 2021. Between 2000 and 2011, regional per capita income grew at about 2.0 percent per year but slowed to 0.4 percent from 2012 to 2019. The poor economic performance led to social protests in several countries, despite their different ideological orientations. This weakened democratic governance, causing a full breakdown in some cases, and contributed to the emergence of authoritarian governments and mass migrations in countries such as Venezuela. The index of democracy, calculated by the Economist, dropped more than 5 percent for the region between 2008 and 2021.[2]

Thus, when the COVID-19 pandemic hit in 2020, many countries in LAC were struggling with both low economic growth and weakened governments. Countries responded to the pandemic with restrictions on mobility and a range of health and income support measures, financed by fiscal and monetary expansion. As a result, the LAC region experienced the largest increase in the debt-to-GDP ratio among developing regions (it rose from 68.4 percent in 2019 to 77.8 percent in 2021 for LAC's group of emerging markets and middle-income developing countries).[3] Notwithstanding the pandemic policy responses, the region, with only about 8 percent of the world's population, suffered about 30 percent of global

FIGURE 1 GDP growth per capita and real price of commodities

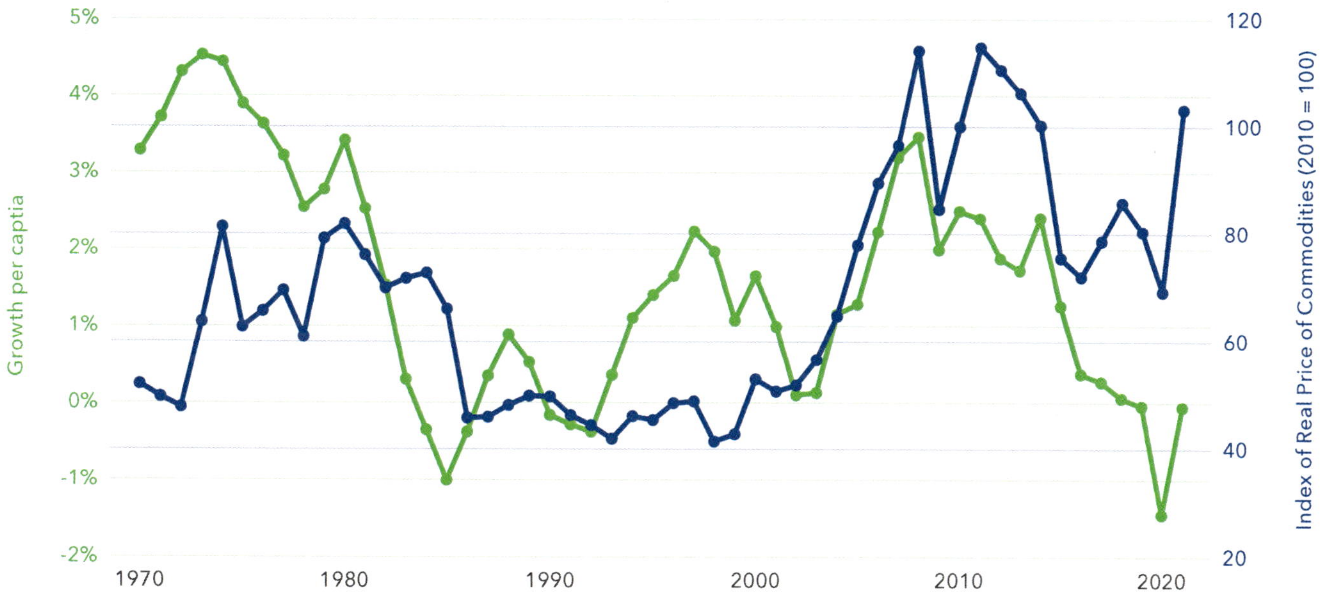

Source: Based on data from the World Bank's World Development Indicators and Commodity Prices database.

deaths. Income per capita fell by about 7.5 percent, more than any other developing region. LAC was particularly vulnerable to the pandemic for several reasons, including its high level of urbanization, significant income inequality (which also limits access to high-quality health services), the informality of labor markets, the prevalence of obesity, and the economic stagnation that preceded the pandemic.[4]

Agricultural production (including forestry and fishing) fared better in 2020 than other economic sectors, due in part to government support to the sector and to the fact that food production and distribution were considered essential activities during the pandemic by most countries, and so faced fewer mobility restrictions. But the deep recession in 2020 that affected demand, plus several climatic events (from hurricanes in Central America to droughts in South America), kept sectoral growth low. Agriculture sector growth was only 0.5 percent in 2020 and 1.2 percent in 2021, compared to more than 3.0 percent worldwide in both years.[5]

In 2021, the region enjoyed a strong economic rebound (up 5.8 percent over 2020), but GDP per capita remained below pre-pandemic levels. At the same time, prices of many products were increasing due to a significant acceleration of world growth (2021 saw the highest per capita growth of any year in the period since 1960), the lingering effects of COVID-19 in logistics chains, and the effects of heatwaves and droughts in a number of agricultural countries. Then, in February 2022, the Russian invasion of Ukraine caused a further surge in the prices of food, fertilizers, and energy. Although those prices have declined since the initial shock, they remain above pre-pandemic levels.

Poverty and nutrition indicators followed the path of the most recent commodity cycle, and then were affected by the pandemic and the war. Economic growth in the upward phase of the commodity cycle, supported by the expansion of social assistance policies in LAC, helped to reduce the percentage of poor people (at US$3.65 PPP/capita/day) from almost 27 percent of the population in 2000 to about 11 percent in 2014-2015. However, the poverty rate stagnated until 2019, when growth declined in the downcycle.[6] Although processed household survey data is not available for all countries in LAC after 2019, extrapolation from those countries with data suggests that the pandemic

FIGURE 2 Undernourishment across LAC, 2000–2021

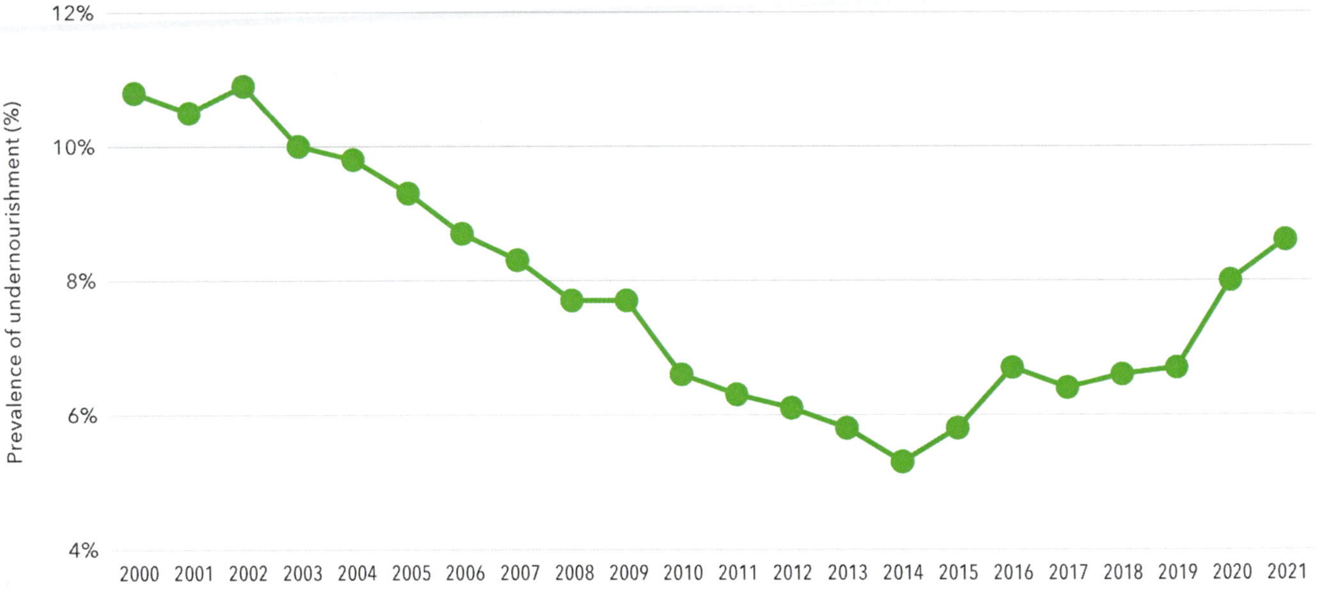

Source: Data from FAOSTAT (2022).

may have raised the regional poverty rate to 14 percent in 2020. Because of the closing of the economy and the nature of many women's jobs in the service sector, women were more affected than men by some of the pandemic-related problems – for example, they were 44 percent more likely than male workers to lose their jobs.[7]

Similarly, undernourishment and hunger (lack of calories) tracked the commodity and growth cycle, with rates declining from almost 11.0 percent of the population in 2000 to 5.3 percent in 2014 as the cycle peaked. The hunger rate trended somewhat higher until 2019, when it reached about 6.7 percent, and then rose to 8.0 percent in 2020 and 8.6 percent in 2021 as a result of the pandemic and the related economic slowdown, with the highest rates in Haiti, Central America, and Venezuela. There are no estimates as yet for 2022.

The most recent data also show that about 22.5 percent of LAC's population cannot afford a healthy diet (2020),[8] and indicate a rising trend in obesity and related noncommunicable diseases. As of 2016, 24 percent of the adult population of LAC was obese, close to the 27 percent found in the United States, Canada, and Europe, and well above the world average of 13 percent. Of course, these poverty and malnutrition problems vary across LAC: Haiti and several countries in Central America are more affected by hunger, poverty, and lack of access to healthy diets, but suffer less from obesity, while obesity is more prevalent in countries such as Argentian, Chile, and Uruguay, and different configurations of those problems exist in between.

All LAC countries have been affected by these macroeconomic, political, health, and climatic shocks over the past decade, but the effects have been more devasting for some than others. Haiti in particular has borne the brunt of a calamitous combination of climate and natural disasters with political, economic, social, and health crises in recent decades. In the past two years alone, its president was assassinated, and the island was hit by an extremely damaging earthquake of 7.2 magnitude followed by another of 5.3 magnitude a few months later.[9]

POLICY CONSIDERATIONS

As the current crises play out, their negative impacts on malnutrition and poverty are likely to be aggravated by the tightening of global fiscal and monetary policies, which is leading to a slowdown in the world economy. Moreover, the extreme weather events already inflicting damage in the region are projected to intensify in the near future.[10] To prepare for and address these threats, LAC governments must confront a number of short- and medium-term challenges, bearing in mind that the application of policy responses will have to be fine-tuned to each country in view of the region's complexity and the large variation between lower- and higher-income countries' human, financial, and innovation capacities.

MANAGE FERTILIZER USE. Global fertilizer prices remain high, despite falling from their peak in April 2022. In the short term, special efforts are needed from LAC governments and the private sector to ensure adequate supply and more efficient use of fertilizers, along with a technological shift toward new fertilizers and management practices with lower greenhouse gas emissions. Countries should organize public-private working groups to monitor fertilizer markets and help ensure their supply.

MANAGE COMMODITY AND ECONOMIC CYCLES. While commodity cycles are inevitable, governments need to manage them better by saving in the upcycle to be able to provide economic support in the downcycle. The international community can help by (1) supporting debt relief through improved mechanisms for debt restructurings and write-offs,[11] (2) increasing the capital of multilateral development banks and optimizing the use of their balance sheets, so they can expand lending, and (3) using scarce international development funds more strategically to leverage and mobilize the vast liquidity in global private capital markets, orienting those markets toward larger humanitarian and developmental objectives (for example, by making better use of the Special Drawing Rights issued by the International Monetary Fund[12]).

INVEST IN SCIENCE, TECHNOLOGY, AND INNOVATION TO ADDRESS CLIMATE CHANGE. The increasing frequency of extreme weather events requires a greater investment in science, technology, and innovation to develop and scale up critical measures for adaptation, resilience, and mitigation. Agriculture in general, and particularly in developing countries, is an important part of the solution to climate change, given its triple potential role of reducing emissions through climate-smart practices; contributing to mitigation by capturing CO_2 through more efficient agriculture and landscape management; and increasing sectoral resilience and adapting to worsening climate and weather conditions. Most of the LAC countries should invest more in agricultural R&D, given that current R&D expenditures in many countries fall below the suggested minimum of 1 percent of agricultural GDP. The need for scaled-up investments in science and technology applies to the whole food value chain and the consumer environment as well. In this regard, it has been suggested that investments in science, technology, and innovation should reach at least 1 percent of *all* GDP related to food systems, not just agricultural GDP.[13]

IMPROVE HEALTH SYSTEMS. The pandemic has highlighted the need for more effective health systems. LAC and the rest of the world are adjusting to a situation in which the COVID-19 virus and its variants are endemic. With vaccines, testing, and the development of better treatments, the disease now seems manageable. However, the future will bring new epidemics, which will require not only strengthening LAC's health systems but also improving global surveillance and rapid-response mechanisms.[14] In particular, a "one health" approach to the interaction of human and animal health in food systems, which has been the source of many recent pandemics, must be supported by strong science and technology investments.

(RE)BUILD HUMAN CAPITAL THROUGH SOCIAL SAFETY NETS AND NUTRITION PROGRAMS. Finally, human capital in LAC, as in other developing regions, has been affected by the nutritional problems associated with insufficient and less-healthy diets as well as setbacks caused by the pandemic, including the gap in education for the current generation of students and the weakening of job skills due to long unemployment periods for some working people. All these problems must be addressed to improve welfare and long-term growth. In particular, it will be necessary to scale up and reevaluate the design of social safety nets and nutrition programs in the region, with the goal of reducing the high levels of inequality and increasing resilience to future crises. A promising option for LAC countries is to expand the focus of cash-transfer programs in rural areas to combine social, productive, and environmental dimensions of sustainable development – with a percentage of cash transfers related to poverty levels; another share to cover the additional cost of implementing sustainable adaptation and mitigation technologies; and another for forest, biodiversity, and other ecosystem protection and restoration services. More generally, it would be beneficial to establish a framework for social inclusion, in both rural and urban settings, with multidimensional programs including social safety nets, livelihoods and jobs, and financial inclusion.[15]

Further, given that cash transfers or vouchers are already being used for both temporary humanitarian programs that respond to recurring crises and expanded permanent social assistance programs, integrating those programs into what has been called "shock-responsive social assistance" could offer a way forward (see Chapters 3 and 5). In addition, the series of negative shocks in recent years has led to many humanitarian programs operating on a near-permanent basis. This has created strong networks of institutions – national and international, public and nongovernmental – with significant experience and operational capabilities on the ground that could be integrated within long-term national strategies developed by elected authorities.

CONCLUSION

The LAC region has experienced significant economic and political volatility, exacerbated by extreme climate events and natural disasters. Several policy measures – related to macroeconomic and financial issues, climate change, health, and social interventions – have been recommended here to address the short- and medium-term challenges generated by those shocks. LAC countries are now burdened by pandemic-related increases in debt and face a host of preexisting economic and social problems as well as the threat of climate-related disasters. To tackle the current challenges and prepare for likely future shocks, they will need substantial financial support from international organizations to implement the policies that will put them on a path toward greater stability, equality, and resilience.

Notes

CHAPTER 1

1 FAO (Food and Agriculture Organization of the United Nations), FAOSTAT database.

2 FAO, IFAD (International Fund for Agricultureal Development), UNICEF, and WHO (World Health Organization), *The State of Food Security and Nutrition in the World 2021: Transforming Food Systems for Food Security, Improved Nutrition and Affordable Healthy Diets for All* (Rome: FAO, 2021).

3 Ibid.

4 World Economic Forum, *Global Gender Gap Report 2022* (2022).

5 J. Glauber, M. Hernández, D. Laborde, W. Martin, B. Rice, and R. Vos, "No End in Sight Yet for the Global Food Price Crisis," *IFPRI Blog*, IFPRI, Sept. 27, 2022.

6 J. Glauber and D. Laborde, "How Sanctions on Russia and Belarus Are Impacting Exports of Agricultural Products and Fertilizer," *IFPRI Blog*, IFPRI, Nov. 9, 2022; B. Fonseca, *Fertilizer Outlook: Is History Repeating Itself?* (Rabobank, Nov. 2022).

7 Food Security Information Network (FSIN) and Global Network Against Food Crises (GNAFC), *2022 Global Report on Food Crises: Joint Analysis for Better Decisions* (Rome: 2022).

8 FSIN and GNAFC, *Global Report on Food Crises: Joint Analysis for Better Decisions* (Rome: 2022); FSIN and GNAFC, *Global Report on Food Crises: Mid-Year Update* (Rome: 2022).

9 IPCC (Intergovernmental Panel on Climate Change), *Climate Change 2022: Impacts, Adaptation and Vulnerability. Contribution of Working Group II to the Sixth Assessment Report of the Intergovernmental Panel on Climate Change* (Cambridge: Cambridge University Press, 2022).

10 A. Ortiz-Bobea, T.R. Ault, C.M. Carrillo, et al., "Anthropogenic Climate Change Has Slowed Global Agricultural Productivity Growth," *Nature Climate Change* 11 (2021): 306–312.

11 R. Treisman, "Officials and Volunteers Struggle to Respond to Catastrophic Flooding in Pakistan," *NPR*, Sept. 1, 2022; Relief Web, "Horn of Africa Drought: Humanitarian Update," June 10, 2022.

12 IFPRI, "Projections from IFPRI's IMPACT Model: Climate Change and Food Systems," in *2022 Global Food Policy Report: Climate Change and Food Systems*, 146-161 (Washington, DC: 2022).

13 G. Pacillo, D. Kangogo, I. Madurga-Lopez, V. Villa, A. Belli, and P. Laderach, "Is Climate Exacerbating the Root Causes of Conflict in Mali? A Climate Security Analysis through a Structural Equation Modeling Approach," *Frontiers in Climate* 4 (2022).

14 IDMC (Internal Displacement Monitoring Centre), *Global Report on Internal Displacement 2021* (2021).

15 ICRC (International Committee of the Red Cross), "The ICRC's Call to COP26 to Strengthen Climate Action in Conflict Settings" (Geneva: 2021).

16 FSIN and GNAFC, *Global Report on Food Crises: Mid-Year Update* (Rome: 2022).

17 FSIN and GNAFC, *Global Report on Food Crises: Mid-Year Update* (Rome: 2022); Development Initiatives, *Global Humanitarian Assistance Report 2020* (Wilmington, DE: 2020).

18 S. Khan, "Female Education and Marriage in Pakistan," WIDER Working Paper 2021/115, UNU-WIDER, Helsinki, 2021); R. Jensen and R. Thornton, "Early Female Marriage in the Developing World," *Gender & Development* 11, 2 (2003): 9-19.

19 J. Chen, K. Kosec, and V. Mueller, "Moving to Despair? Migration and Well-Being in Pakistan," *World Development* 113 (2019): 186-203.

20 J. Barchfield, "Pandemic Deepens Hunger for Displaced People the World Over," UNHCR News & Stories, March 2021.

21 IDMC (Internal Displacement Monitoring Centre), *Global Report on Internal Displacement* (Geneva: 2022).

22 M. Hidrobo, N. Kumar, T. Palermo, A. Peterman, and S. Roy, *Gender-Sensitive Social Protection: A Critical Component of the COVID-19 Response in Low- and Middle-Income Countries* (Washington, DC: IFPRI, 2020).

23 K. Mohammed, D. Najjar, and E. Bryan, "Women's Resilience and Participation in Climate Governance in the Agri-food Sector: A Strategic Review of Public Policies," ICARDA Working Paper, ICARDA, Beirut, Dec. 2022; D. Amoak, D. Najjar, and J. Kyle, "Gender and Climate-Relevant Agri-Food Systems Governance: A Strategic Evidence Review," ICARDA Working Paper, ICARDA, Beirut, Dec. 2022.

24 R.S. Meinzen-Dick, L.R. Brown, H.S. Feldstein, and A.R. Quisumbing, "Gender, Property Rights, and Natural Resources, *World Development* 25, 8 (1997): 1303-1315.

25 M. Hidrobo, K. Kosec, H. Gartaula, B. Van Campenhout, and L. Carrillo, "Making Complementary Agricultural Resources, Technologies, and Services More Gender-Responsive," Background paper for *Report on the Status of Rural Women in Agriculture: Gender Equality in Agriculture, Food Security and Nutrition*, under review 2023.

26 UNHCR (United Nations High Commissioner for Refugees), *2021 Global Trends in Forced Displacement* (Geneva: 2020).

27 K. Kosec, J. Song, H. Zhao, and B. Holtemeyer, "The Gendered Impacts of Income Fluctuations on Household Departure, Labor Supply, and Human Capital Decisions: Evidence from Kyrgyzstan," *Feminist Economics* (2022): 1-31.

28 C.R. Doss, A. Qaisrani, K. Kosec, V. Slavchevska, A. Galiè, and N. Kawarazuka, "From the `Feminization of Agriculture' to Gender Equality,"in *Advancing Gender Equality Through Agricultural and Environmental Research: Past, Present, and Future*, eds. R. Pyburn and A. van Eerdewijk, Ch. 8 (Washington, DC: IFPRI, 2021).

29 See IFPRI's Food and Fertilizer Export Restrictions Tracker.

30 CGIAR Research Program on Policies, Institutions, and Markets (PIM), *Understanding Characteristics, Causes, and Consequences of Migration: Contributions from the CGIAR Research Program on Policies, Institutions, and Markets* (Washington, DC: IFPRI, 2021).

31 Carrefour Chief Executive Alexandre Bompard, at the 2022 Consumer Goods Global Summit, https://www.theconsumergoodsforum.com/events/the-global-summit/

32 S. Narayanan and S. Saha, "One Step Behind: The Government of India and Agricultural Policy during the Covid-19 Lockdown," *Review of Agrarian Studies* 10, 1 (2020): 111-127; T. Reardon, J. Swinnen, and R. Vos, "COVID-19 and Resilience Innovations in Food Supply Chains: Two Years Later," in *COVID-19 and Global Food Security: Two Years Later*, eds. J. McDermott and J. Swinnen, 87-91 (Washington, DC: IFPRI, 2022).

33 J.E. Taylor, M.J. Filipski, M. Alloush, A. Gupta, R.I.R. Valdes, and E. Gonzalez-Estrada, "Economic Impact of Refugees," *PNAS* 113, 27 (2016): 7449–7453.

34 K. Kosec, and D. Resnick, "Governance: Making Institutions Work for Rural Revitalization," in *2019 Global Food Policy Report*, Ch. 8 (Washington, DC: IFPRI, 2019).

35 D.K. Evans, B. Holtemeyer, and K. Kosec, "Cash Transfers Increase Trust in Local Government," *World Development* 114 (2019): 138–155; K. Kosec and C.H. Mo, "Does Relative Deprivation Condition the Effects of Social Protection Programs on Political Support? Experimental Evidence from Pakistan," *American Journal of Political Science*, forthcoming.

36 R.D. Kusumastuti, Z.A. H. Viverita, L. Suardi, and D.N. Danarsari, "Developing a Resilience Index toward Natural Disasters in Indonesia," *International Journal of Disaster Risk Reduction* 10 (2014): 327–340.

37 M. Ahmed, S.N. Nawal, and P. Bhangle, "Impact of Good Governance on Disaster Risk Management," in *Peace, Justice and Strong Institutions*, eds. W. Leal Filho, A.M. Azul, L. Brandli, A. Lange Salvia, P.G. Özuyar, and T. Wall (Cham, Switzerland: Springer, 2021).

38 K. Kosec and L. Wantchekon, "Can Information Improve Rural Governance and Service Delivery?" *World Development* 125 (2020): 104376.

39 Humanitarian Action, "At a Glance," Nov. 30, 2022.

40 U. Gentilini, M. Almenfi, H. TMM Iyengar, et al., "Tracking Global Social Protection Responses to Price Shocks: Living Paper," version 1, Social Protection & Jobs Discussion Paper 2208, World Bank, Washington, DC, 2022.

41 World Bank, *Food Finance Architecture : Financing a Healthy, Equitable, and Sustainable Food System* (Washington, DC: 2021).

42 R. Vos, W. Martin, and D. Resnick, "Repurposing Agricultural Support: Creating Food Systems Incentives to Address Climate Change," in *2022 Global Food Policy Report: Climate Change and Food Systems*, Ch. 2 (Washington, DC: IFPRI).

43 BSDC (Business and Sustainable Development Commission), *Valuing the SDG Prize in Food and Agriculture: Unlocking Business Opportunities to Accelerate Sustainable and Inclusive Growth* (London: 2016).

44 World Bank, *Food Finance Architecture : Financing a Healthy, Equitable, and Sustainable Food System* (Washington, DC: 2021).

45 M. Rowling, "Explainer: What's the Plan to Fill the Climate Finance Gap for Poor Nations?," Reuters, Nov. 10, 2022.

46 World Bank Databank, Population Estimates and Projections, accessed January 23, 2023.

BOX 1

1 J. Glauber, M. Hernández, D. Laborde, W. Martin, B. Rice, and R. Vos, "No End in Sight Yet for the Global Food Price Crisis," *IFPRI Blog*, IFPRI, Sept. 27 2022.

2 C. Arndt, X. Diao, P. Dorosh, K. Pauw, and J. Thurlow, *Russia-Ukraine War and the Global Food Crisis: Impacts on Poverty and Food Security in Developing Countries*, IFPRI Country Brief 20 (Washington, DC: IFPRI, 2022).

3 K. Abay, N. Karachiwalla, S. Kurdi, and Y. Salama, "Food Pirce Shocks and Diets among Poor Households in Egypt," *IFPRI Blog*, IFPRI, Dec. 29, 2022.

4 R. Vos, J. Glauber, and D. Laborde, "Is Food Price Inflation Really Subsiding?" *IFPRI Blog*, IFPRI, Jan. 24, 2023.

CHAPTER 2

1 World Food Programme (WFP), WFP Global Operational Response Plan: Update #6, Nov. 2022 (Rome: 2022).

2 C.P. Timmer, "Preventing Food Crises Using a Food Policy Approach," *Journal of Nutrition* 140, 1 (2009): 2245–2285.

3 Food Security Information Network (FSIN) and Global Network Against Food Crises (GNAFC), *2022 Global Report on Food Crises: Mid-Year Update* (Rome: 2022).

4 J. Glauber, M.A. Hernandez, D. Laborde, W. Martin, B. Rice, and R. Vos, "No End in Sight Yet for the Global Food Price Crisis," *IFPRI Blog*, Sept. 27, 2022.

5 FSIN and GNAFC, *2022 Global Report on Food Crises: Mid-Year Update* (Rome: 2022).

6 FSIN and GNAFC, *2022 Global Report on Food Crises: Mid-Year Update* (Rome: 2022).

7 G7 (Group of Seven), *G7 Principles Supporting a Vision of Improved Global Food Security Monitoring and Analysis* (London: UK Foreign Affairs, 2021).

8 IPC (Integrated Food Security Phase Collection) Global Partners, *Technical Manual Version 3: Evidence and Standards for Better Food Security and Nutrition Decisions* (Rome: 2021).

9 FEWS NET (Famine Early Warning Systems Network), *Scenario Development for Food Security Early Warning*, Guidance Document No. 1 (Washington, DC: 2018).

10 HungerMap^{LIVE}, https://hungermap.wfp.org/

11 WFP, *Technical Guidance for WFP: Consolidated Approach for Reporting Indicators of Food Security (CARI)*, third ed. (Rome: 2021).

12 WFP and Food and Agriculture Organization (FAO), *Hunger Hotspots: FAO-WFP Early Warnings on Acute Food Insecurity*, October 2022 to January 2023 Outlook (Rome: WFP, 2022.)

13 Both measures are part of the official indicators for monitoring progress toward SDG2 (ending hunger and all forms of malnutrition) and reported annually in the report *The State of Food Security and Nutrition in the World*. The FIES is a relatively new and second-tier indicator and not yet officially adopted by all countries.

14 Agriculture Market Information System (AMIS), https://www.amis-outlook.org/amis-about/en/

15 GEOGLAM Crop Monitor, https://cropmonitor.org/

16 CropWatch, http://www.cropwatch.com.cn/htm/en/about.shtml

17 C. Holleman, J. Jackson, M.V. Sánchez, and R. Vos, *Sowing the Seeds of Peace for Food Security: Disentangling the Nexus between Conflict, Food Security and Peace*, FAO Agricultural Development Economics Technical Study 2 (Rome: FAO, 2017); WFP USA, *Winning the Peace: Hunger and Instability*, (Washington, DC: 2017); SIPRI (Stockholm International Peace Research Institute), Food, Peace and Security, accessed January 2023.

18 W. Martin and N. Minot, "The Impacts of Price Insulation on World Wheat Markets during the 2022 Food Price Crisis," *Agricultural and Resource Economics* 6, 4 (2022): 753–774.

19 P. Foini, M. Tizzoni, G. Martini, et al. "On the Forecastability of Food Insecurity," *Scientific Reports* 13, 2793 (2023).

20 B. Andrée, A. Chamorro, A. Kraay, P. Spencer, and D. Wang, "Predicting Food Crises," Policy Research Working Paper 9412, World Bank, Washington, DC, 2020.

21 World Bank Shock Responsive Safety Net for Human Capital Project, https://projects.worldbank.org/en/projects-operations/project-detail/P171346

22 World Bank, "Colombia's Cat DDO: Strengthening Resilience to Disasters, Climate Change and Health Risks," World Bank Treasury Case Study (Washington, DC: 2022).

23 IFRC (International Federation of Red Cross and Red Crescent Societies) and German Red Cross, "Forecast-based Financing," accessed January 2023. https://www.forecast-based-financing.org/

24 WFP, *Forecast-Based Financing (FbF): Anticipatory Actions for Food Security* (Rome: 2019).

25 WFP, "Hunger Monitoring. Tracking Hunger in Real Time," Factsheet (Rome: 2022); G. Martini, A. Bracci, L. Riches, et al. "Machine Learning Can Guide Food Security Efforts When Primary Data Are Not Available," *Nature Food* 3 (2022): 716-728.

26 Y. Zhou, E. Lentz, H. Michelson, C. Kim, and K. Baylis, "Machine Learning for Food Security: Principles for Transparency and Usability," *Applied Economic Perspectives and Policy* 44, 2 (2021): 893-910; L. McBride, C.B. Barrett, et al. "Predicting Poverty and Malnutrition for Targeting, Mapping, Monitoring, and Early Warning," *Applied Economic Perspectives and Policy* 44, 2 (2021): 879-892.

27 G20 (Group of 20), "G20 Bali Leaders' Declaration," G20 Bali Summit, November 15-16, 2022; G7, "G7 Statement on Global Food Security," G7 Germany, June 28, 2022.

BOX 1

1 CFS (Committee on World Food Security), "Reform of the Committee on World Food Security," CFS:2009/2 Rev.2 (Rome: FAO, 2009).

2 FAO (Food and Agriculture Organization of the United Nations), IFAD (International Fund for Agricultural Development), UNICEF, WFP (World Food Programme), and WHO (World Health Organization), *The State of Food Security and Nutrition in the World 2022: Repurposing Food and Agricultural Policies to Make Healthy Diets More Affordable* (Rome: FAO, 2022).

3 FSIN (Food Security Information Network) and GNAFC (Global Network Against Food Crises), *2022 Global Report on Food Crises: Joint Analysis for Better Decisions* (Rome: 2022); IPC (Integrated Food Security Phase Collection) Global Partners, *Technical Manual Version 3: Evidence and Standards for Better Food Security and Nutrition Decisions* (Rome: 2021).

4 IPC Global Partners, *Technical Manual Version 3: Evidence and Standards for Better Food Security and Nutrition Decisions* (Rome: 2021).

BOX 2

1 IPC (Integrated Food Security Phase Collection) Global Partners, *Technical Manual Version 3: Evidence and Standards for Better Food Security and Nutrition Decisions* (Rome: 2021).

CHAPTER 3

1 Aon, *Weather, Climate, and Catastrophe* (London: 2021).

2 J. Glauber and D. Laborde, eds. *The Russia-Ukraine Conflict and Global Food Security* (Washington, DC: IFPRI, forthcoming).

3 E. Skoufias, "Economic Crises and Natural Disasters: Coping Strategies and Policy Implications," *World Development* 31, 7 (2003): 1087-1102.

4 OCHA (United Nations Office for the Coordination of Humanitarian Affairs), *Global Humanitarian Review 2022* (New York: 2021).

5 J.H.S. Lie, "The Humanitarian-Development Nexus: Humanitarian Principles, Practice, and Pragmatics," *Journal of International Humanitarian Action* 5, (2020): 18.

6 D. Maxwell, L. Russo, and L. Alinovi, "Constraints to Addressing Food Insecurity in Protracted Crises," *Proceedings of the National Academy of Sciences* 109, 31 (2012): 12321-12325.

7 IFPRI, "Social Protection by Design" (Washington, DC: 2015).

8 J. Puri, A. Aladysheva, V. Iversen, Y. Ghorpade, and T. Brück, "Can Rigorous Impact Evaluations Improve Humanitarian Assistance?" *Journal of Development Effectiveness* 9, 4 (2017): 519-542.

9 D. Jeong and I. Trako, *Cash and In-Kind Transfers in Humanitarian Settings: A Review of Evidence and Knowledge Gaps* (Washington, DC: World Bank, 2022).

10 A. Banerjee, E. Duflo, N. Goldberg, et al. , "A Multifaceted Program Causes Lasting Progress for the Very Poor: Evidence from Six Countries," *Science* 348, 6236 (2015): 1260799; L. Brune, D. Karlan, S. Kurdi, and C. Udry, "Social Protection amidst Social Upheaval: Examining the Impact of a Multi-faceted Program for Ultra-Poor Households in Yemen," *Journal of Development Economics* 155 (2022): 102780.

11 J.P. Tranchant, A. Gelli, L. Bliznashka, et al., "The Impact of Food Assistance on Food Insecure Populations during Conflict: Evidence from a Quasi-Experiment in Mali," *World Development* 119 (2019): 185-202.

12 E. Aurino, J.P. Tranchant, A. Sekou Diallo, and A. Gelli, "School Feeding or General Food Distribution? Quasi-Experimental Evidence on the Educational Impacts of Emergency Food Assistance during Conflict in Mali," *Journal of Development Studies* 55, S1 (2019): 7-28.

13 Global Nutrition Cluster, *Evidence and Guidance Note on the Use of Cash and Voucher Assistance for Nutrition Outcomes in Emergencies* (New York: UNICEF, 2022).

14 M. Hidrobo, J. Hoddinott, A. Peterman, A. Margolies, and V. Moreira, "Cash, Food, or Vouchers? Evidence from a Randomized Experiment in Northern Ecuador," *Journal of Development Economics* 107 (2014): 144-156; J. Hoddinott, S. Sandström, and J. Upton, "The Impact of Cash and Food Transfers: Evidence from a Randomized Intervention in Niger," *American Journal of Agricultural Economics* 100, 4 (2018): 1032-1049; D.O. Gilligan, M. Hidrobo, J. Hoddinott, et al., *Expanding Lessons from a Randomised Impact Evaluation of Cash and Food Transfers in Ecuador and Uganda*, 3ie Grantee Final Report (New Delhi: 3ie, 2015); B. Schwab, "In the Form of Bread? A Randomized Comparison of Cash and Food Transfers in Yemen," *American Journal of Agricultural Economics* 102, 1 (2020): 91-113.

15 S. Kurdi, "The Nutritional Benefits of Cash Transfers in Humanitarian Crises: Evidence from Yemen," *World Development* 148 (2021): 105664; O. Ecker, J-F. Maystadt, and Z. Guo, "Can Unconditional Cash Transfers Mitigate the Impact of Civil Conflict on Acute Child Malnutrition in Yemen? Evidence from the National Social Protection Monitoring Survey," MENA Regional Program Working Paper 17, IFPRI, Washington, DC, 2019.

16 IAHE (Inter-Agency Humanitarian Evaluation), *Interagency Humanitarian Evaluation of the Yemen Crisis* (New York: UN OCHA, 2022); IAHE, *Inter-Agency Humanitarian Evaluation of the Drought Response in Ethiopia 2015-2018* (New York: UN OCHA, 2020).

17 D.K. Evans, B. Holtemeyer, and K. Kosec, "Cash Transfers Increase Trust in Local Government," *World Development* 114 (2019): 138-155.

18 R. Sabates-Wheeler, K. Hirvonen, J. Lind, and J. Hoddinott, "Expanding Social Protection Coverage with Humanitarian Aid: Lessons on

Targeting and Transfer Values from Ethiopia," *Journal of Development Studies* 58, 10 (2022): 1981–2000.

19 G. Dafuleya, "Social and Emergency Assistance Ex-Ante and During COVID-19 in the SADC Region," *International Journal of Community and Social Development* 2, 2 (2020): 251–268.

20 P. Howe, "The Triple Nexus: A Potential Approach to Supporting the Achievement of the Sustainable Development Goals?" *World Development* 124 (2019): 104629.

21 L. Huybregts, F. Houngbé, C. Salpéteur, R. Brown, D. Roberfroid, M. Ait-Aissa, and P. Kolsteren, "The Effect of Adding Ready-to-Use Supplementary Food to a General Food Distribution on Child Nutritional Status and Morbidity: A Cluster-Randomized Controlled Trial," *PLoS Medicine* 9, 9 (2012): e1001313; M.T. Ruel, H. Alderman, and Maternal and Child Nutrition Study Group, "Nutrition-Sensitive Interventions and Programmes: How Can They Help to Accelerate Progress in Improving Maternal and Child Nutrition?" *Lancet* 382, 9891 (2013): 536–551.

22 D.K. Olney, Q. Marshall, G. Honton, et al., "Leveraging an Implementation-Research Partnership to Improve Effectiveness of Nutrition-Sensitive Programs at the World Food Programme," *Food and Nutrition Bulletin* 41, 1 (2020): 18–37.

23 C. Del Ninno, P.A. Dorosh, and K. Subbarao, "Food Aid, Domestic Policy and Food Security: Contrasting Experiences from South Asia and Sub-Saharan Africa," *Food Policy* 32, 4 (2007): 413–435.

24 A.M. Kizito and E. Kato, "Does Linking Farmers to Markets Work? Evidence from the World Food Programme's Purchase for Progress Satellite Collection Points Initiative in Uganda," *African Journal of Agricultural and Resource Economics* 13, 2 (2018): 169–181

25 L. Zanotti, "Cacophonies of Aid, Failed State Building and NGOs in Haiti: Setting the Stage for Disaster, Envisioning the Future," *Third World Quarterly* 31, 5 (2010): 755–771; S. Campbell, M. DiGiuseppe, and A. Murdie, "International Development NGOs and Bureaucratic Capacity: Facilitator or Destroyer?" *Political Research Quarterly* 72, 1 (2019): 3–18.

26 U. Gentilini, S. Laughton, and C. O'Brien, "Human(itarian) Capital? Lessons on Better Connecting Humanitarian Assistance and Social Protection," Social Protection and Jobs Discussion Paper No. 1802, World Bank, Washington, DC, 2018.

27 S. Levine, E. Wilkinson, and L. Weingärtner, "Anticipatory Action for Livelihood Protection: A Collective Endeavour," Working Paper No. 580, Overseas Development Institute (ODI), London, 2020.

28 WFP (World Food Programme), *Forecast-Based Financing (FbF): Anticipatory Actions for Food Security* (Rome: 2019).

29 A. Seal and R. Bailey, "The 2011 Famine in Somalia: Lessons Learnt from a Failed Response?" *Conflict and Health* 7 (2013): 22; D. Maxwell, K. Gelsdorf, N. Haan, D. Dawe, eds., "Special Issue on the Somalia Famine of 2011–2012," Special issue, *Global Food Security* 1, 1 (December 2012).

30 WMO (World Meteorological Organization), *WMO Guidelines on Multi-hazard Impact-based Forecast and Warning Services: Part II: Putting Multi-hazard IBFWS into Practice* (Geneva: 2021).

31 P. Raschky and M. Schwindt, "Aid, Natural Disasters and the Samaritan's Dilemma," World Bank Policy Research Working Paper No. 4951, World Bank, Washington, DC, 2009.

32 S. Levine, E. Wilkinson, and L. Weingärtner, "Anticipatory Action for Livelihood Protection: A Collective Endeavour," Working Paper No. 580, ODI, London, 2020.

33 AATF (Anticipatory Action Task Force), "Enabling Anticipatory Action at Scale," Policy Brief for Donor Governments, 2021.

34 FAO (Food and Agriculture Organization of the United Nations), "Anticipatory Action," presented at the Food Security & Livelihoods Cluster Coordination meeting, Rome, May 21, 2021.

35 IFRC (International Federation of Red Cross and Red Crescent Societies), *2021 Annual Plan: Disaster Relief Emergency Fund and Forecast-based Action* (Geneva: 2020).

36 D. MacLeod, D.R. Kniveton, and M.C. Todd, "Playing the Long Game: Anticipatory Action Based on Seasonal Forecasts," *Climate Risk Management* 34 (2021): 100375.

37 L. Weingärtner and E. Wilkinson, "Anticipatory Crisis Financing and Action: Concepts, Initiatives and Evidence," Working Paper, ODI, London, 2019.

38 P. Venton, C.C. Venton, N. Limones, et al., "Framework for the Assessment of Benefits of Action/Cost of Inaction (BACI) for Drought Preparedness," Water Global Practice Working Paper, World Bank Group, Washington, DC, 2019.

39 C. Gros, E. Easton-Calabria, M. Bailey, K. Dagys, E.C. de Perez, M. Sharavnyambuu, and A. Kruczkiewicz, "The Effectiveness of Forecast-Based Humanitarian Assistance in Anticipation of Extreme Winters: A Case Study of Vulnerable Herders in Mongolia," *Disasters* 46, 1 (2022): 95–118.

40 C. Gros, M. Bailey, S. Schwager, et al., "Household-Level Effects of Providing Forecast-Based Cash in Anticipation of Extreme Weather Events: Quasi-Experimental Evidence from Humanitarian Interventions in the 2017 Floods in Bangladesh," *International Journal of Disaster Risk Reduction* 41 (2019): 101275.

41 See https://www.ifpri.org/topic/resilience for a repository of publications on this topic.

CHAPTER 4

1 T. Reardon and B. Minten, "Food Value Chain Transformation in Developing Regions," in *Agricultural Development: New Perspectives in a Changing World*, eds. K. Otsuka and S. Fan, 397–437 (Oxford: Oxford University Press, 2020).

2 J. von Braun and E. Kennedy, *Commercialization of Subsistence Agriculture: Income and Nutritional Effects in Developing Countries* (Washington, DC: IFPRI, 1986).

3 T. Reardon and D. Zilberman, "Climate Smart Food Supply Chains in Developing Countries in an Era of Rapid Dual Change in Agrifood Systems and the Climate," in *Climate Smart Agriculture*, eds. L. Lipper, N. McCarthy, D. Zilberman, S. Asfaw, and G. Branca, 335–351 (Cham, Switzerland: Springer, 2018).

4 Ibid.

5 L.S.O. Liverpool-Tasie, T. Reardon, and B. Belton, "'Essential Non-essentials': COVID-19 Policy Missteps in Nigeria Rooted in Persistent Myths about African Food Supply Chains," *Applied Economic Perspectives and Policy* 43, 1 (2020): 205–224; T. Reardon, A. Mishra, C.S.R. Nuthalapati, M. Bellemare, and D. Zilberman, "COVID-19's Disruption of India's Transformed Food Supply Chains," *Economic and Political Weekly* 55, 18 (2020): 7–8.

6 L.S.O. Liverpool-Tasie, B. Belton, O. Tasie, et al., *Pivoting in Nigeria's Fish and Poultry Value Chains in Response to COVID-19 Policies and Impacts*, NAPA Research Paper 1 (East Lansing: Michigan State University, 2021).

7 Ibid.

8 B. Belton, L. Rosen, L. Middleton, et al., "COVID-19 Impacts and Adaptations in Asia and Africa's Aquatic Food Value Chains," *Marine Policy* 129 (2021): 104523.

9 L.S.O. Liverpool-Tasie, B. Belton, O. Tasie, et al., *Pivoting in Nigeria's Fish and Poultry Value Chains in Response to COVID-19 Policies and Impacts*, NAPA Research Paper 1 (East Lansing: Michigan State University, 2021).

10 D. Naziri, B. Belton, S. Alobo-Loison, T. Reardon, et al., "Fish and Chips in Crisis! COVID-19 Disruptions and Pivoting in the Hidden Middle of Kenya's Potato and Fish Value Chains," forthcoming.

11 S. Nordhagen, U. Igbeka, H. Rowlands, R.S. Shine, E. Heneghan, and J. Tench, "COVID-19 and Small Enterprises in the Food Supply Chain: Early Impacts and Implications for Longer-Term Food System Resilience in Low- and Middle-Income Countries," *World Development* 141 (2021): 105405.

12 C. Béné, D. Bakker, M.J. Chavarro, B. Even, J. Melo, and A. Sonneveld, "Global Assessment of the Impacts of COVID-19 on Food Security," *Global Food Security* 31 (2021): 100575.

13 K. Van Hoyweghen, A. Fabry, H. Feyaerts, I. Wade, and M. Maertens, "Resilience of Global and Local Value Chains to the COVID-19 Pandemic: Survey Evidence from Vegetable Value Chains in Senegal," *Agricultural Economics* 52, 3 (2021): 423–440.

14 B. Minten, B. Mohammed, and S. Tamru, "Emerging Medium-Scale Tenant Farming, Gig Economies, and the COVID-19 Disruption: The Case of Commercial Vegetable Clusters in Ethiopia," *European Journal of Development Research* 32, 5 (2020): 1402–1429.

15 D. Naziri, B. Belton, S. Alobo-Loison, T. Reardon, et al., "Fish and Chips in Crisis! COVID-19 Disruptions and Pivoting in the Hidden Middle of Kenya's Potato and Fish Value Chains," forthcoming

16 HLPE (High Level Panel of Experts on Food Security and Nutrition), *Impacts of COVID-19 on Food Security and Nutrition: Developing Effective Policy Responses to Address the Hunger and Malnutrition Pandemic* (Rome: FAO, 2020).

17 K. Hirvonen, B. Minten, B. Mohammed, and S. Tamru, "Food Prices and Marketing Margins during the COVID19 Pandemic: Evidence from Vegetables in Ethiopia," *Agricultural Economics* 52, 3 (2021): 407–421.

18 I. Ali, A. Arslan, M. Chowdhury, Z. Khan, and S.Y. Tarba, "Reimagining Global Food Value Chains through Effective Resilience to COVID-19 Shocks and Similar Future Events: A Dynamic Capability Perspective," *Journal of Business Research*, 141 (2022): 1–12.

19 D.C. Love, E.H. Allison, F. Asche, et al., "Emerging COVID-19 Impacts, Responses, and Lessons for Building Resilience in the Seafood System," *Global Food Security* 28 (2021): 100494.

20 D. Naziri, B. Belton, S. Alobo-Loison, T. Reardon, et al., "Fish and Chips in Crisis! COVID-19 Disruptions and Pivoting in the Hidden Middle of Kenya's Potato and Fish Value Chains," forthcoming.

21 T. Reardon, A. Heiman, L. Lu, C.S.R. Nuthalapati, R. Vos, and D. Zilberman, "'Pivoting' by Food Industry Firms to Cope with COVID-19 in Developing Regions: E-commerce and 'Copivoting' Delivery Intermediaries," *Agricultural Economics* 52, 3 (2021): 459–475.

22 D. Naziri, B. Belton, S. Alobo-Loison, T. Reardon, et al., "Fish and Chips in Crisis! COVID-19 Disruptions and Pivoting in the Hidden Middle of Kenya's Potato and Fish Value Chains," forthcoming.

23 L.S.O. Liverpool-Tasie, B. Belton, O. Tasie, et al., *Pivoting in Nigeria's Fish and Poultry Value Chains in Response to COVID-19 Policies and Impacts*, NAPA Research Paper 1 (East Lansing: Michigan State University, 2021).

24 B. Belton, *COVID-19 and Lessons Learned on the Resilience of Aquaculture Value Chains* (Punchong, Malaysia: INFOFISH International, 2022).

25 T. Reardon, A. Heiman, L. Lu, C.S.R. Nuthalapati, R. Vos, and D. Zilberman, "'Pivoting' by Food Industry Firms to Cope with COVID-19 in Developing Regions: E-commerce and 'Copivoting' Delivery Intermediaries," *Agricultural Economics* 52, 3 (2021): 459–475.

26 D. Naziri, B. Belton, S. Alobo-Loison, T. Reardon, et al., "Fish and Chips in Crisis! COVID-19 Disruptions and Pivoting in the Hidden Middle of Kenya's Potato and Fish Value Chains," forthcoming.

27 OECD, *States of Fragility 2020* (Paris: 2020).

28 C. Hebebrand and D. Laborde, "High Fertilizer Prices Contribute to Rising Global Food Security Concerns," *IFPRI Blog*, April 22, 2022.

29 Ibid.

30 M. Ruta, ed., *The Impact of the War in Ukraine on Global Trade and Investment* (Washington, DC: World Bank, 2022).

31 Ibid; D. Laborde, W. Martin, J. Swinnen, and R. Vos, "COVID-19 Risks to Global Food Security," *Science* 369, 6503 (2020): 500–502.

32 W. Martin and N. Minot, "The Impacts of Price Insulation on World Wheat Markets during the 2022 Food Price Crisis," *Australian Journal of Agricultural and Resource Economics* 66, 4 (2022): 753–774.

33 World Bank, *Myanmar Economic Monitor*, January 2022 (Washington, DC: 2022); MAPSA (Myanmar Agriculture Policy Support Activity), "Double Jeopardy: COVID-19, Coup d'État and Poverty in Myanmar," Myanmar Strategy Support Program Working Paper 22, IFPRI, Washington, DC, 2022.

34 B. Minten, J. Goeb, K.Z. Win, and P. Zone, "Agricultural Value Chains in a Fragile State: Evidence from Rice in Myanmar," MAPSA Discussion Paper 15, IFPRI, Washington, DC, 2022.

35 Ibid.

36 C. Del Ninno, P.A. Dorosh, and L.C. Smith, "Public Policy, Markets and Household Coping Strategies in Bangladesh: Avoiding a Food Security Crisis following the 1998 Floods," *World Development* 31, 7 (2003): 1221–1238.

37 C. Murendo, A. Keil, and M. Zeller, "Drought Impacts and Related Risk Management by Smallholder Farmers in Developing Countries: Evidence from Awash River Basin, Ethiopia," *Risk Management* 13, 4 (2011): 247–263.

38 B. Kramer, P. Hazell, H. Alderman, F. Ceballos, N. Kumar, and A.G. Timu, "Is Agricultural Insurance Fulfilling its Promise for the Developing World?" IFPRI Discussion Paper 2057, IFPRI, Washington, DC, 2021.

39 M. Carter, A. de Janvry, E. Sadoulet, and A. Sarris, "Index Insurance for Developing Country Agriculture: A Reassessment," *Annual Review of Resource Economics* 9, 1(2017): 421–438.

40 B. Kramer, P. Hazell, H. Alderman, F. Ceballos, N. Kumar, and A.G. Timu, "Is Agricultural Insurance Fulfilling Its Promise for the Developing World?" IFPRI Discussion Paper 2057, IFPRI, Washington, DC, 2021; see also Chapter 5 in this report.

41 D. Gollin, *Impact of International Research on Intertemporal Yield Stability in Wheat and Maize: An Economic Assessment* (Mexico City: CIMMYT, 2006).

42 C.B. Barrett, T. Reardon, J. Swinnen, and D. Zilberman, "Agri-food Value Chain Revolutions in Low- and Middle-Income Countries," *Journal of Economic Literature* 60, 4 (2022): 1316–1377; N.J. Sitko and T.S. Jayne, "Why are African Commodity Exchanges Languishing? A Case Study of the Zambian Agricultural Commodity Exchange," *Food Policy* 37, 3 (2012): 275–282.

43 S. Jaffee, P. Siegel, and C. Andrews, "Rapid Agricultural Supply Chain Risk Assessment: A Conceptual Framework," Agriculture and Rural Development Discussion Paper 47, World Bank, Washington, DC, 2010.

44 D. Boughton, J. Goeb, I. Lambrecht, et al., "Impacts of COVID-19 on Agricultural Production and Food Systems in Late Transforming

Southeast Asia: The Case of Myanmar," *Agricultural Systems* 188 (2021): 103026.

45 World Food Programme, *The WFP's Real-Time Monitoring System: Approaches and Methodologies* (Rome: 2021).

CHAPTER 5

1 S. Hickey, T. Lavers, M. Niño-Zarazúa, and J. Seekings, "The Negotiated Politics of Social Protection in Sub-Saharan Africa," UNU-WIDER Working Paper 2018/34, UNU-WIDER, Helsinki, 2018.

2 K. Beegle, M. Honorati, and E. Monsalve, "Reaching the Poor and Vulnerable in Africa through Social Safety Nets," in *Realizing the Full Potential of Social Safety Nets in Africa*, eds. K. Beegle, A. Coudouel, and E. Monsalve, 49-87 (Washington, DC: World Bank, 2018).

3 M. Hidrobo, J. Hoddinott, N. Kumar, and M. Olivier, "Social Protection, Food Security, and Asset Formation," *World Development* 101 (2018): 88-103; C. Andrews, A. Hsiao, and L. Ralston, "Social Safety Nets Promote Poverty Reduction, Increase Resilience, and Expand Opportunities," in *Realizing the Full Potential of Social Safety Nets in Africa*, eds. K. Beegle, A. Coudouel, and E. Monsalve, 87-138 (Washington, DC: World Bank, 2018).

4 D.K. Olney, A. Gelli, N. Kumar, H. Alderman, A. Go, and A. Raza, "Social Assistance Programme Impacts on Women's and Children's Diets and Nutritional Status," *Maternal & Child Nutrition* 18, 4 (2022): e13378; J. Manley, Y. Balarajan, S. Malm, et al., "Cash Transfers and Child Nutritional Outcomes: A Systematic Review and Meta-Analysis," *BMJ Global Health* 5 (2020): e003621; J. Manley, H. Alderman, and U. Gentilini, "More Evidence on Cash Transfers and Child Nutritional Outcomes: A Systematic Review and Meta-Analysis," *BMJ Global Health* 7 (2022): e008233.

5 K. Kosec and C.H. Mo, "Aspirations and the Role of Social Protection: Evidence from a Natural Disaster in Rural Pakistan," *World Development* 97 (2017): 49-66.

6 D.O. Gilligan, S. Devereux, and J. Tenzing, "Social Protection: Designing Adaptive Systems to Build Resilience to Climate Change," in *2022 Global Food Policy Report: Climate Change & Food Systems*, 58-63 (Washington, DC: IFPRI, 2022).

7 T. Fetzer, "Can Workfare Programs Moderate Conflict? Evidence from India," *Journal of the European Economic Association* 18 (2020): 3337-3375.

8 D.K. Evans, B. Holtemeyer, and K. Kosec, "Cash Transfers Increase Trust in Local Government," *World Development* 114 (2019): 138-155; K. Kosec and C.H. Mo, "Does Relative Deprivation Condition the Effects of Social Protection Programs on Political Attitudes? Experimental Evidence from Pakistan," *American Journal of Political Science*, forthcoming.

9 H. Alderman and R. Yemtsov, "How Can Safety Nets Contribute to Economic Growth?" *World Bank Economic Review* 28 (2014):1-20; M. Filipski, J.E. Taylor, G.A. Abegaz, T. Ferede, A.S. Taffesse, and X. Diao, "Synopsis: Economy-wide impacts of the Productive Safety Net Programme (PSNP)," ESSP II Resarch Note 57 (Washington, DC: IFPRI and Ethiopian Development Research Institute, 2016).

10 K. Ambler and A. De Brauw, "The Impacts of Cash Transfers on Women's Empowerment: Learning from Pakistan's BISP Program," Social Protection and Labor Discussion Paper No. 1702, World Bank, Washington, DC, 2017.

11 A.M. Buller, A. Peterman, M. Ranganathan, A. Bleile, M. Hidrobo, and L. Heise, "A Mixed-Method Review of Cash Transfers and Intimate Partner Violence in Low- and Middle-Income Countries," *World Bank Research Observer* 33 (2018): 218-258; R. Heath, M. Hidrobo, and S. Roy, "Cash Transfers, Polygamy, and Intimate Partner Violence: Experimental Evidence from Mali," *Journal of Development Economics* 143 (2020): 102410; S. Roy, M. Hidrobo, J. Hoddinott, and A. Ahmed, "Transfers, Behavior Change Communication, and Intimate Partner Violence: Postprogram Evidence from Rural Bangladesh," *Review of Economics and Statistics* 101, 5 (2019): 865-877.

12 A. Castañeda, D. Doan, D. Newhouse, M.C. Nguyen, H. Uematsu, and J.P. Azevedo, "A New Profile of the Global Poor," *World Development* 101 (2018): 250-267.

13 O. Hoegh-Guldberg, D. Jacob, M. Taylor, et al., "Impacts of 1.5°C Global Warming on Natural and Human Systems," in *Special Report: Global Warming of 1.5°C*, Ch. 3 (Geneva: IPCC, 2018).

14 World Bank, *Poverty and Shared Prosperity 2022: Correcting Course* (Washington, DC, 2022); D. Headey and K. Hirvonen, "Food Inflation, Poverty and Urbanization," IFPRI Discussion Paper 2134, IFPRI, Washington, DC, 2022.

15 U. Gentilini, S. Khosla, and M. Almenfi, "Cash in the City: Emerging Lessons from Implementing Cash Transfers in Urban Africa," Social Protection & Jobs Discussion Paper No. 2101, World Bank, Washington, DC, 2021; U. Gentilini, *Cash Transfers in Pandemic Times: Evidence, Practices, and Implications from the Largest Scale Up in History* (Washington, DC: World Bank, 2022); J. Duchoslav and K. Hirvonen, "Delivery of Social Protection Programs to Combat COVID-19 in Africa," in *ReSAKKS Annual Trends and Outlook Report: Building Resilient African Food Systems after COVID-19*, eds. J. Ulimwengy, M.A. Constas, and E. Ubalijoro, 100-111 (Kigali and Washington, DC: AKADEMIYA2063 and IFPRI, 2021).

16 U. Gentilini, *Cash Transfers in Pandemic Times* (Washington, DC: World Bank, 2022); U. Gentilini, M.B.A. Almenfi, T. Iyengar, et al., *Social Protection and Jobs Responses to COVID-19* (Washington, DC: World Bank, 2022).

17 A. McCord, "The Impact of the Global Financial Crisis on Social Protection in Developing Countries," *International Social Security Review* 63 (2010): 31-45.

18 M. Grosh, C. Andrews, R. Quintana, and C. Rodriguez-Alas, "Assessing Safety Net Readiness in Response to Food Price Volatility," Social Protection & Jobs Discussion Paper No. 1118, World Bank, Washington, DC, 2011; World Bank, *The World Bank Group and the Global Food Crisis: An Evaluation of the World Bank Group Response* (Washington, DC: 2014).

19 World Bank, *The State of Social Safety Nets 2018* (Washington, DC: 2018); S. Hickey, T. Lavers, M. Niño-Zarazúa, and J. Seekings, "The Negotiated Politics of Social Protection in Sub-Saharan Africa," UNU-WIDER Working Paper 2018/34, UNU-WIDER, Helsinki, 2018.

20 World Bank, ASPIRE (Atlas of Social Protection: Indicators of Resilience and Equity) database, June 28, 2022, update.

21 M. Hidrobo, J. Hoddinott, N. Kumar, and M. Olivier, "Social Protection, Food Security, and Asset Formation," *World Development* 101 (2018): 88-103.

22 K. Beegle, M. Honorati, and E. Monsalve, "Reaching the Poor and Vulnerable in Africa through Social Safety Nets," in *Realizing the Full Potential of Social Safety Nets in Africa*, eds. K. Beegle, A. Coudouel, and E. Monsalve, 49-87 (Washington, DC: World Bank, 2018).

23 E. Knippenberg and J. Hoddinott, "Shocks, Social Protection, and Resilience: Evidence from Ethiopia," ESSP Working Paper 109, IFPRI, Addis Ababa, 2017.

24 K. Haile, "Cash Transfers, Negative Rainfall Shocks and Child Welfare in Ethiopia," *Journal of African Economies* 31, 5 (2021): 441-466.

25 S. Asfaw, A. Carraro, B. Davis, S. Handa, and D. Seidenfeld, "Cash Transfer Programmes, Weather Shocks and Household Welfare: Evidence from a Randomised Experiment in Zambia," *Journal of Development Effectiveness* 9 (2017): 419–442.

26 P. Premand and Q. Stoeffler, "Cash Transfers, Climatic Shocks and Resilience in the Sahel," *Journal of Environmental Economics and Management* 116 (2022): 102744.

27 X. Hou, "Can Drought Increase Total Calorie Availability? The Impact of Drought on Food Consumption and the Mitigating Effects of a Conditional Cash Transfer Program," *Economic Development and Cultural Change* 58 (2010): 713–737.

28 A. De Janvry, F. Finan, E. Sadoulet, and R. Vakis, "Can Conditional Cash Transfer Programs Serve as Safety Nets in Keeping Children at School and from Working when Exposed to Shocks?" *Journal of Development Economics* 79 (2006): 349–373.

29 A. Adhvaryu, A. Nyshadham, T. Molina, and J. Tamayo, "Helping Children Catch Up: Early Life Shocks and the Progresa Experiment," NBER Working Paper No. 24848, National Bureau of Economic Research, Cambridge, MA, 2018.

30 P. Christian, E. Kandpal, N. Palaniswamy, and V. Rao, "Safety Nets and Natural Disaster Mitigation: Evidence from Cyclone Phailin in Odisha," *Climatic Change* 153 (2019): 141–164.

31 O. Ivaschenko, J. Doyle, J. Kim, J. Sibley, and Z. Majoka, "Does 'Manna from Heaven' Help? The Role of Cash Transfers in Disaster Recovery—Lessons from Fiji after Tropical Cyclone Winston," *Disasters* 44 (2020): 455–476.

32 J. McDermott and J. Swinnen, *COVID-19 and Global Food Security: Two Years Later* (Washington, DC: IFPRI, 2022).

33 J. Londoño-Vélez and P. Querubin, "The Impact of Emergency Cash Assistance in a Pandemic: Experimental Evidence from Colombia," paper presented at the 113th Annual Conference on Taxation, Washington, DC, November 18–20, 2020.

34 N. Bottan, B. Hoffmann, and D.A. Vera-Cossio, "Stepping Up during a Crisis: The Unintended Effects of a Noncontributory Pension Program during the Covid-19 Pandemic," *Journal of Development Economics* 150 (2021): 102635.

35 K.A. Abay, G. Berhane, J.F. Hoddinott, and K. Tafere, "COVID-19 and Food Security in Ethiopia: Do Social Protection Programs Protect?" *Economic Development and Cultural Change* 71, 2 (2023): 372–402.

36 A. Banerjee, M. Faye, A. Krueger, P. Niehaus, and T. Suri, "Effects of a Universal Basic Income during the Pandemic," Working Paper, Innovations for Poverty Action, New York, 2020.

37 W. Brooks, K. Donovan, T.R. Johnson, and J. Oluoch-Aridi, "Cash Transfers as a Response to COVID-19: Experimental Evidence from Kenya," *Journal of Development Economics* 158 (2021): 102929.

38 H. Alderman, J. Hoddinott, and B. Kinsey, "Long Term Consequences of Early Childhood Malnutrition," *Oxford Economic Papers-New Series* 58 (2006): 450–474; S. Maccini and D. Yang, "Under the Weather: Health, Schooling, and Economic Consequences of Early-Life Rainfall," *American Economic Review* 99 (2009): 1006–1026; S. Dercon and C. Porter. 2014. "Live Aid Revisited: Long-Term Impacts of the 1984 Ethiopian Famine on Children," *Journal of the European Economic Association* 12 (2014): 927–948.

39 A. Urquhart, F. Girling, E. Mason, and S. Nelson-Pollard, *Global Humanitarian Assistance Report 2022* (Wilmington, DE: Development Initiatives, 2022).

40 C. Cherrier, "Cash Transfers and Resilience: Strengthening Linkages between Emergency Cash Transfers and National Social Transfer Programmes in the Sahel," Discussion Paper, CALP Network, Oxford, UK, 2014.

41 UN OCHA (United Nations Office for the Coordination of Humanitarian Affairs), *The Grand Bargain: A Shared Commitment to Better Serve People in Need* (Istanbul: 2016)

42 C. Cherrier, "Cash Transfers and Resilience," Discussion Paper, CALP Network, Oxford, UK, 2014.

43 U. Gentilini, *Cash Transfers in Pandemic Times* (Washington, DC: World Bank, 2022); K.A. Abay, N. Yonzan, S. Kurdi, and K. Tafere, *Revisiting Poverty Trends and the Role of Social Protection Systems in Africa during the COVID-19 Pandemic* (Washington, DC: World Bank, 2022).

44 C. Cherrier, "Cash Transfers and Resilience," Discussion Paper, CALP Network, Oxford, UK, 2014; M. Ulrichs and R. Sabates-Wheeler, "Social Protection and Humanitarian Response?" IDS Working Paper 516, Institute of Development Studies, Brighton, UK, 2018.

45 T. Bowen, C. del Ninno, C. Andrews et al., *Overview: A Framework for Adaptive Social Protection* (Washington, DC: World Bank, 2020).

46 A. Pople, P. Premand, S. Dercon, T. Bossuroy, S. Brunelin, and M. Vinez, "What is the Value of Early Response? Multi-country RCT of Shock-Responsive Cash Transfers in the Sahel," AEA RCT Registry, 2022.

47 M. Farhat, F. Merttens, and C. Riungu, *Evaluation of the Kenya Hunger Safety Net Programme Phase 2: Emergency Payments Deep Dive Study* (Oxford: Oxford Policy Management, 2017).

48 T. Bowen, C. del Ninno, C. Andrews, et al., *Overview: A Framework for Adaptive Social Protection* (Washington, DC: World Bank, 2020).

49 W. Wiseman, J. Van Domelen, and S. Coll-Black, *Designing and Implementing a Rural Safety Net in a Low Income Setting: Lessons Learned from Ethiopia's Productive Safety Net Program 2005-2009* (Washington, DC: World Bank, 2010).

50 K. Beegle, M. Honorati, and E. Monsalve, "Reaching the Poor and Vulnerable in Africa through Social Safety Nets," in *Realizing the Full Potential of Social Safety Nets in Africa*, eds. K. Beegle, A. Coudouel, and E. Monsalve, 49–87 (Washington, DC: World Bank, 2018).

51 G. Berhane, D.O. Gilligan, J. Hoddinott, N. Kumar, and A.S. Taffesse, "Can Social Protection Work in Africa? The Impact of Ethiopia's Productive Safety Net Programme," *Economic Development and Cultural Change* 63 (214): 1–26; E. Knippenberg and J. Hoddinott, "Shocks, Social Protection, and Resilience: Evidence from Ethiopia, ESSP Working Paper 109, IFPRI, Addis Ababa, 2017.

52 R. Sabates-Wheeler, K. Hirvonen, J. Lind, and J. Hoddinott, "Expanding Social Protection Coverage with Humanitarian Aid: Lessons on Targeting and Transfer Values from Ethiopia," *Journal of Development Studies* 58, 10 (2022): 1981–2000.

53 NDRMC (National Disaster Risk Management Commission), *Ethiopia: Humanitarian and Disaster Resilience Plan 2018, Mid-Year Review* (Addis Ababa: Joint Government and Humanitarian Partners' Document, NDRMC, 2018).

54 World Bank, *Quantifying Costs of Drought Risk in Ethiopia: A Technical Note* (Washington, DC: 2017); World Bank, *Ethiopia Poverty Assessment of Social Protection: Indicators of Resilience and Equity* (Washington, DC: 2020); K. Endale, A. Pick, and T. Woldehanna, *Financing Social Protection in Ethiopia: A Long-Term Perspective*, Development Policy Papers (Paris: OECD, 2019).

55 Development Initiatives, *Global Humanitarian Assistance Report 2012* (Wilmington, DE: 2012).

56 R. Sabates-Wheeler, K. Hirvonen, J. Lind, and J. Hoddinott, "Expanding Social Protection Coverage with Humanitarian Aid: Lessons on

Targeting and Transfer Values from Ethiopia," *Journal of Development Studies* 58, 10 (2022): 1981-2000.

57 T. Bowen, C. del Ninno, C. Andrews, et al., *Overview: A Framework for Adaptive Social Protection* (Washington, DC: World Bank, 2020); F. Bastagli, *Responding to a Crisis: The Design and Delivery of Social Protection* (London: Overseas Development Institute, 2014).

58 T. Bowen, C. del Ninno, C. Andrews, et al., *Overview: A Framework for Adaptive Social Protection* (Washington, DC: World Bank, 2020).

59 A. Pople, R. Hill, S. Dercon, and B. Brunckhorst, "Anticipatory Cash Transfers in Climate Disaster Response," Working Paper 6, Centre for Disaster Protection, London, 2021.

60 A. Tohari, C. Parsons, and A. Rammohan, "Targeting Poverty under Complementarities: Evidence from Indonesia's Unified Targeting System," *Journal of Development Economics* 140 (2019): 127-144.

61 UNICEF, UNDP (United Nations Development Programme), Prospera, and SMERU Research Institute, *Analysis of the Social and Economic Impacts of COVID-19 on Households and Strategic Policy Recommendations for Indonesia* (Jakarta: 2021).

62 T. Bowen, C. del Ninno, C. Andrews, et al. , *Overview: A Framework for Adaptive Social Protection* (Washington, DC: World Bank, 2020).

63 D. Eckstein, V. Künzel, L. Schäfer, and M. Winges, *Global Climate Risk Index 2020: Who Suffers Most from Extreme Weather Events? Weather-Related Loss Events in 2018 and 1999 to 2018* (Bonn: Germanwatch, 2019).

64 T. Bowen, C. del Ninno, C. Andrews, et al., *Overview: A Framework for Adaptive Social Protection* (Washington, DC: World Bank, 2020); D.J. Clarke and S. Dercon, *Dull Disasters? How Planning Ahead Will Make a Difference* (Oxford: Oxford University Press, 2016).

65 D.J. Clarke and S. Dercon, *Dull Disasters? How Planning Ahead Will Make a Difference* (Oxford: Oxford University Press, 2016).

66 T. Bowen, C. del Ninno, C. Andrews, et al., *Overview: A Framework for Adaptive Social Protection* (Washington, DC: World Bank, 2020).

67 World Bank, *The State of Social Safety Nets 2018* (Washington, DC: 2018).

68 K. Hirvonen, E. Machado, A. Simons, and V. Taraz, "More than a Safety Net: Ethiopia's Flagship Public Works Program Increases Tree Cover in Rural Areas and Sloped Terrains," *Global Environmental Change* 75 (2022): 102549; P.J. Ferraro and R. Simorangkir, "Conditional Cash Transfers to Alleviate Poverty also Reduced Deforestation in Indonesia," *Science Advances* 6, 24 (2020): eaaz1298.

69 M. Tefera, A. Strickland, J. Lind, D. Woolf, S. Jirka, and D. Solomon, *Carbon Finance Opportunities for Ethiopia's PSNP*, CSI Policy Brief 6 (Ithaca, NY: Cornell University, 2015); S. Jirka, D. Woolf, D. Solomon, and J. Lehmann, *Climate Finance and Carbon Markets for Ethiopia's Productive Safety Net Programme (PSNP): Executive Summary for Policymakers* (Ithaca, NY: Cornell University, 2015); E. Díaz-Bonilla and R.G. Echeverría, "Climate Finance: Funding Sustainable Food Systems Transformation," in *2022 Global Food Policy Report: Climate Change and Food Systems*, Ch. 5, 48-57 (Washington, DC: IFPRI, 2022).

70 J.C. Aker, R. Boumnijel, A. McClelland, and N. Tierney, "Payment Mechanisms and antipoverty Programs: Evidence from a Mobile Money Cash Transfer Experiment in Niger," *Economic Development and Cultural Change* 65 (2016): 1-37; M. Hidrobo, J. Hoddinott, A. Peterman, A. Margolies, and V. Moreira, "Cash, Food, or Vouchers? Evidence from A Randomized Experiment in Northern Ecuador," *Journal of Development Economics* 107 (2014): 144-156.

71 K. Hirvonen and J. Hoddinott, "Beneficiary Views on Cash and In-Kind Payments: Evidence from Ethiopia's Productive Safety Net Programme," *World Bank Economic Review* 35 (2021): 398-413; L. Gadenne, S. Norris, M. Singhal, and S. Sukhtankar, "In-Kind Transfers as Insurance," NBER Working Paper No. 28507, National Bureau of Economic Research, Cambridge MA, 2021; R. Sabates-Wheeler and S. Devereux, "Cash Transfers and High Food Prices: Explaining Outcomes on Ethiopia's Productive Safety Net Programme," *Food Policy* 35 (2010): 274-285.

72 A. Banerjee, E. Duflo, N. Goldberg, et al., "A Multifaceted Program Causes Lasting Progress for the Very Poor: Evidence from Six Countries," *Science* 348, 6236 (2015): 1260799.

BOX 1

1 World Bank, *The State of Social Safety Nets 2018* (Washington, DC: 2018).

2 N. Parekh and O. Bandiera, *Do Social Assistance Programmes Reach the Poor? Micro-evidence from 123 Countries*, IGC Growth Brief Series 023. (London: International Growth Centre, 2020).

3 H. Alderman, U. Gentilini, and R. Yemtsov, *The 1.5 Billion People Question: Food, Vouchers, or Cash Transfers?* (Washington, DC: World Bank, 2017).

4 D. Coady, M. Grosh, and J. Hoddinott, "Targeting Outcomes Redux," *World Bank Research Observer* 19, 1 (2004): 61-85.

5 N. Benhassine, F. Devoto, E. Duflo, P. Dupas, and V. Pouliquen, "Turning a Shove into a Nudge? A 'Labeled Cash Transfer' for Education," *American Economic Journal: Economic Policy* 7, 3 (2015): 86-125; R. Akresh, D. De Walque, and H. Kazianga, "Evidence from a Randomized Evaluation of the Household Welfare Impacts of Conditional and Unconditional Cash Transfers Given to Mothers or Fathers," Policy Research Working Paper No. 7730, World Bank, Washington, DC, 2016; J. Haushofer and J. Shapiro, "The Short-Term Impact of Unconditional Cash Transfers to the Poor: Experimental Evidence from Kenya," *Quarterly Journal of Economics* 131 (2016): 1973-2042.

6 U. Gentilini, M. Grosh, J. Rigolini, and R. Yemtsov, *Exploring Universal Basic Income: A Guide to Navigating Concepts, Evidence, and Practices* (Washington, DC: World Bank, 2019).

BOX 2

1 A. Banerjee, E. Duflo, N. Goldberg, et al., "A Multifaceted Program Causes Lasting Progress for the Very Poor: Evidence from Six Countries," *Science* 348, 6236 (2015): 1260799.

2 A. Banerjee, E. Duflo, and G. Sharma, "Long-Term Effects of the Targeting the Ultra Poor Program," *American Economic Review: Insights* 3, 4 (2021): 471-486.

3 O. Bandiera, R. Burgess, N. Das, S. Gulesci, I. Rasul, and M. Sulaiman, "Labor Markets and Poverty in Village Economies," *Quarterly Journal of Economics* 132, 2 (2017): 811-870; C. Balboni, O. Bandiera, R. Burgess, M. Ghatak, and A. Heil, "Why Do People Stay Poor?" *Quarterly Journal of Economics* 137, 2 (2022):785-844.

4 L. Brune, D. Karlan, S. Kurdi, and C. Udry, "Social Protection amidst Social Upheaval: Examining the Impact of a Multi-faceted Program for Ultra-Poor Households in Yemen," *Journal of Development Economics* 155 (2022): 102780; G. Bedoya, A. Coville, J. Haushofer, M. Isaqzadeh, and J.P. Shapiro, "No Household Left Behind: Afghanistan Targeting the Ultra Poor Impact Evaluation," NBER Working Paper No. 25981, National Bureau of Economic Research, Cambridge, MA, 2019; A. Banerjee, D. Karlan, R. Osei, H. Trachtman, and C. Udry, "Unpacking a Multi-faceted Program to Build Sustainable Income for the Very Poor," *Journal of Development Economics* 155 (2022): 102781.

5 T. Bossuroy, M. Goldstein, B. Karimou, et al., "Tackling Psychosocial and Capital Constraints to Alleviate Poverty," *Nature* 605, 7909 (2022): 291-297.

CHAPTER 6

1 OECD (Organisation for Economic Co-operation and Development), *States of Fragility 2022* (Paris: 2022).

2 UNDP (United Nations Development Program), *Human Development Report 2021-22. Uncertain Times, Unsettled Lives: Shaping Our Future in a Transforming World* (New York: 2022).

3 OECD, *States of Fragility 2022* (Paris: 2022).

4 WEF (World Economic Forum), *Global Gender Gap Report 2022* (Cologny, Switzerland: 2022).

5 OECD, *States of Fragility 2022* (Paris: 2022).

6 V.M. Hudson, B. Ballif-Spanvill, M. Caprioli, and C.F. Emmett, *Sex and World Peace* (New York: Columbia University Press, 2014); V.M. Hudson, "What Sex Means for World Peace," *Foreign Policy*, April 24, 2012.

7 D.L. Bowen, V.M. Hudson, and P.L. Nielsen, "State Fragility and Structural Gender Inequality in Family Law: An Empirical Investigation," *Laws* 4, 4 (2015): 654-672.

8 M. Caprioli, "Primed for Violence: The Role of Gender Inequality in Predicting Internal Conflict," *International Studies Quarterly* 49 (2005): 161-178; M. Caprioli and M.A. Boyer, "Gender, Violence, and International Crisis," *Journal of Conflict Resolution* 45, 4 (2016): 503-518; J.H.R. Demeritt, A.D. Nichols, and E.G. Kelly, "Female Participation and Civil War Relapse," *Civil Wars* 16 (2015): 346-368; E. Melander, "Gender Equality and Intrastate Armed Conflict," *International Studies Quarterly* 49, 4 (2005): 695-714.

9 OECD, *How Fragile Contexts Affect the Well-Being and Potential of Women and Girls* (Paris: 2022).

10 M. Buvinic, M. das Gupta, U. Casabonne, and P. Verwimp, "Violent Conflict and Gender Inequality: An Overview," *World Bank Research Observer* 28, 1 (2013): 110-138.

11 The United Nations Office for Disaster Risk Reduction (UNDRR) defines disaster as "a serious disruption of the functioning of a community or a society at any scale due to hazardous events interacting with conditions of exposure, vulnerability and capacity, leading to one or more of the following: human, material, economic and environmental losses and impacts." https://www.undrr.org/terminology/disaster

12 N. Kumar and A.R. Quisumbing, "Gendered Impacts of the 2007-2008 Food Price Crisis: Evidence Using Panel Data from Rural Ethiopia," *Food Policy* 38 (2013): 11-22; A.R. Quisumbing, N. Kumar, and J.A. Behrman, "Do Shocks Affect Men's and Women's Assets Differently? Evidence from Bangladesh and Uganda," *Development Policy Review* 36 (2018): 3-34; H.J. Brinkman, S. de Pee, I. Sanogo, L. Subran, and M.W. Bloem, "High Food Prices and the Global Financial Crisis Have Reduced Access to Nutritious Food and Worsened Nutritional Status and Health," *Journal of Nutrition* 140, 1 (2010): 153S-161S (2010).

13 S. Fisseha, G. Sen, T.A. Ghebreyesus, et al., "COVID-19: The Turning Point for Gender Equality," *Lancet* 398, 10299 (2021): 471-474; A. Kalbarczyk, N-L. Aberman, B.S.M. van Asperen, et al., "COVID-19, Nutrition, and Gender: An Evidence-Informed Approach to Gender-Responsive Policies and Programs," *Social Science & Medicine* 312 (2022): 115364; C. de Paz Nieves, I. Gaddis, and M. Muller, "Gender and COVID-19: What Have We Learnt, One Year Later?" Policy Research Working Paper No. 9709, World Bank, Washington, DC, 2021; E. Bryan, C. Ringler, and N. Lefore, "To Ease the World Food Crisis, Focus Resources on Women and Girls," *Nature* 609 (2022): 28-31.

14 M. Buvinic, M. das Gupta, U. Casabonne, and P. Verwimp, "Violent Conflict and Gender Inequality: An Overview," *World Bank Research Observer* 28, 1 (2013): 110-138.

15 A.M. Thurston, H. Stöckl, and M. Ranganathan, "Natural Hazards, Disasters and Violence against Women and Girls: A Global Mixed-Methods Systematic Review," *BMJ Global Health* 6, 4 (2021): e004377; L. Stark, I. Seff, and C. Reis, "Gender-Based Violence against Adolescent Girls in Humanitarian Settings: A Review of the Evidence," *Lancet Child & Adolescent Health* 5, 3 (2021): 210-222; A. Vu, A. Adam, A. Wirtz, et al., "The Prevalence of Sexual Violence among Female Refugees in Complex Humanitarian Emergencies: A Systematic Review and Meta-analysis," *PLoS Currents* 6 (2014); K.R. van Daalen, S.S. Kallesøe, F. Davey, et al., "Extreme Events and Gender-Based Violence: A Mixed-Methods Systematic Review," *Lancet Planetary Health* 6,6 (2022): E504-E523.

16 J. Njuki, S. Eissler, H. Malapit, R. Meinzen-Dick, E. Bryan, and A. Quisumbing, "A Review of Evidence on Gender Equality, Women's Empowerment, and Food Systems," *Global Food Security* 33 (2022): 100622.

17 K.A. Abay, T. Kibrom, G. Berhane, J. Chamberlin, and M.H. Abay, "Near-Real-Time Welfare and Livelihood Impacts of an Active Civil War: Evidence from Ethiopia," IFPRI Discussion Paper 2108, IFPRI, Washington, DC, 2022.

18 K. Abay, M.H. Abay, G. Berhane, J. Chamberlin, K. Croke, and K. Tafere, "Access to Health Services, Food, and Water during an Active Conflict: Evidence from Ethiopia," *PLOS Global Public Health*, forthcoming.

19 S. Gourlay, T. Kilic, A. Martuscelli, P. Wollburg, and A. Zezza, "Viewpoint: High-Frequency Phone Surveys on COVID-19: Good Practices, Open Questions," *Food Policy* 105 (2021): 102153.

20 G. Ferrant and A. Thim, *Measuring Women's Economic Empowerment: Time Use Data and Gender Inequality* (Paris: OECD, 2019).

21 WHO (World Health Organization), *Violence Against Women Prevalence Estimates, 2018 Global, Regional and National Prevalence Estimates for Intimate Partner Violence Against Women and Global and Regional Prevalence Estimates for Non-partner Sexual Violence against Women* (Geneva: 2021).

22 The World Bank maintains a global tracker of studies on COVID-19 and violence against women and girls (VAWG), see H. Brixi, H. Fu, and P. Uribe, "Global Crisis of Violence against Women and Girls: Tackling It with New, Better Data Use," *World Bank Data Blog*, Jan. 5, 2022.

23 J. True, "Gendered Violence in Natural Disasters: Learning from New Orleans, Haiti and Christchurch," *Aotearoa New Zealand Social Work* 25 (2013): 78-89.

24 A.M. Thurston, H. Stöckl, and M. Ranganathan, "Natural Hazards, Disasters and Violence Against Women and Girls: A Global Mixed-Methods Systematic Review," *BMJ Global Health* 6, 4 (2021): e004377; R. van Daalen, S.S. Kallesøe, F. Davey, et al., "Extreme Events and Gender-Based Violence: A Mixed-Methods Systematic Review," *Lancet Planetary Health* 6, 6 (2022): E504-E523; L. Stark, I. Seff, and C. Reis, "Gender-Based Violence against Adolescent Girls in Humanitarian Settings: A Review of the Evidence," *Lancet Child & Adolescent Health* 5, 3 (2021): 210-222.

25 L. Stark, I. Seff, and C. Reis, "Gender-Based Violence against Adolescent Girls in Humanitarian Settings: A Review of the Evidence," *Lancet Child & Adolescent Health* 5, 3 (2021): 210-222; J.C. Campbell, "Health Consequences of Intimate Partner Violence," *Lancet* 359, 9314 (2002): 1331-133.

26 A.M. Thurston, H. Stöckl, and M. Ranganathan, "Natural Hazards, Disasters and Violence against Women and Girls: A Global Mixed-Methods Systematic Review," *BMJ Global Health* 6, 4 (2021): e004377.

27 Ibid.

28 Brennan and Nandy define complex humanitarian emergencies as "humanitarian crises characterized by political instability, armed conflict, large population displacements, food shortages, social disruption and collapse of public health infrastructure"; R.J. Brennan and R. Nandy, "Complex Humanitarian Emergencies: A Major Global Health Challenge," *Emergency Medicine* 13, 2 (2001): 147-156.

29 Sexual violence outcomes included rape, molestation, sexual abuse, genital mutilation, gang rape, marital rape, sexual violence related to exploitation, and sexual harassment.

30 A.M. Thurston, H. Stöckl, and M. Ranganathan, "Natural Hazards, Disasters and Violence Against Women and Girls: A Global Mixed-Methods Systematic Review," *BMJ Global Health* 6, 4 (2021): e004377; L. Stark, I. Seff, and C. Reis, "Gender-Based Violence against Adolescent Girls in Humanitarian Settings: A Review of the Evidence," *Lancet Child & Adolescent Health* 5, 3 (2021): 210-222; A. Vu, A. Adam, A. Wirtz, et al., "The Prevalence of Sexual Violence among Female Refugees in Complex Humanitarian Emergencies: A Systematic Review and Meta-Analysis," *PLoS Currents* 6 (2014).

31 L. Stark, I. Seff, and C. Reis, "Gender-Based Violence against Adolescent Girls in Humanitarian Settings: A Review of the Evidence," *Lancet Child & Adolescent Health* 5, 3 (2021): 210-222.

32 J. Austin, S. Guy, L. Lee-Jones, T. McGinn, and J. Schlecht, "Reproductive Health: A Right for Refugees and Internally Displaced Persons," *Reproductive Health Matters* 16, 31 (2008): 10-21.

33 E.O. Akpan, "Early Marriage in Eastern Nigeria and the Health Consequences of Vesicovaginal Fistulae (VVF) among Young Mothers," *Gender & Development* 11, 2 (2003): 70-76; O. Adedokun, O. Adeyemi, and C. Dauda, "Child Marriage and Maternal Health Risks among Young Mothers in Gombi, Adamawa State, Nigeria: Implications for Mortality, Entitlements and Freedoms," *African Health Sciences* 16, 4 (2016): 986-999; K.G. Santhya, U. Ram, R. Acharya, S.J. Jejeebhoy, F. Ram, and A. Singh, "Associations between Early Marriage and Young Women's Marital and Reproductive Health Outcomes: Evidence from India," *International Perspectives on Sexual and Reproductive Health* 36, 3 (2010): 132-139; A. Raj, N. Saggurti, D. Balaiah, and J.G. Silverman, "Prevalence of Child Marriage and Its Effect on Fertility and Fertility-Control Outcomes of Young Women in India: A Cross-Sectional, Observational Study," *Lancet* 373, 9678 (2009): 1883-1889; S. Fan and A. Koski, "The Health Consequences of Child Marriage: A Systematic Review of the Evidence," *BMC Public Health* 22 (2022): 309.

34 S.R. Psaki, A.J. Melnikas, E. Haque, et al., "What Are the Drivers of Child Marriage? A Conceptual Framework to Guide Policies and Programs," *Journal of Adolescent Health* 69, 6S (2021): S13-S22; UNICEF, "Child Marriage on the Rise in Horn of Africa as Drought Crisis Intensifies," press release, June 28, 2022.

35 E. Bryan, C. Ringler, and N. Lefore, "To Ease the World Food Crisis, Focus Resources on Women and Girls," *Nature* 609 (2022): 28-31; UNICEF, "Child Marriage on the Rise in Horn of Africa as Drought Crisis Intensifies," press release, June 28, 2022.

36 M. Alvi, S. Gupta, P. Barooah, C. Ringler, E. Bryan, and R.S. Meinzen-Dick, *Gendered Impacts of COVID-19: Insights from 7 Countries in Sub-Saharan Africa and South Asia* (Washington, DC: IFPRI, 2022).

37 M. Alvi, S. Gupta, P. Barooah, C. Ringler, E. Bryan, and R.S. Meinzen-Dick, *Gendered Impacts of COVID-19: Insights from 7 Countries in Sub-Saharan Africa and South Asia* (Washington, DC: IFPRI, 2022).

38 ALNAP (Active Learning Network for Accountability and Performance), *The State of the Humanitarian System* (London: 2022).

39 Development Initiatives, *Global Humanitarian Assistance Report* (Wilmington, DE: 2022).

40 ALNAP, *The State of the Humanitarian System* (London: 2022); Development Initiatives, *Global Humanitarian Assistance Report* (Wilmington, DE: 2022).

41 Development Initiatives, *Global Humanitarian Assistance Report* (Wilmington, DE: 2022).

42 E. Lwamba, S. Shisler, W. Ridlehoover, et al., *Strengthening Women's Empowerment and Gender Equality in Fragile Contexts towards Peaceful and Inclusive Societies*, 3ie Systematic Review 47 (New Delhi: 3ie, 2021).

43 Ibid.

44 M. Bizzarri, M.C. Vinas, H. Khogali, and T. Jantzi, *Inter-Agency Humanitarian Evaluation on Gender Equality and the Empowerment of Women and Girls* (New York: UN OCHA, 2020).

45 M. Buvinic, M. das Gupta, U. Casabonne, and P. Verwimp, "Violent Conflict and Gender Inequality: An Overview," *World Bank Research Observer* 28, 1 (2013): 110-138.

46 IDMC (Internal Displacement Monitoring Centre), *Global Report on Internal Displacement: Part 2 Children and Youth in Internal Displacement* (Geneva: 2022); IDMC, IMPACT, and Plan International, *Women and Girls in Internal Displacement* (Geneva: 2020).

47 Ibid.

48 A.R. Quisumbing, N. Kumar, and J.A. Behrman, "Do Shocks Affect Men's and Women's Assets Differently? Evidence from Bangladesh and Uganda," *Development Policy Review* 36 (2018): 3-34; E. Bryan, C. Ringler, and N. Lefore, "To Ease the World Food Crisis, Focus Resources on Women and Girls," *Nature* 609 (2022): 28-31.

49 L. Stark, I. Seff, and C. Reis, "Gender-Based Violence against Adolescent Girls in Humanitarian Settings: A Review of the Evidence," *Lancet Child & Adolescent Health* 5, 3 (2021): 210-222.

50 WFP (World Food Programme), "GSMA and UN World Food Programme Accelerate the Use of Mobile Financial Services for Humanitarian Assistance," *WFP News Releases*, August 7, 2020; WFP, *Digital Financial Inclusion Through Cash Transfers: Her Money, Her Account* (Rome: 2022).

51 M. Alvi, S. Gupta, P. Barooah, C. Ringler, E. Bryan, and R.S. Meinzen-Dick, *Gendered Impacts of COVID-19: Insights from 7 Countries in Sub-Saharan Africa and South Asia* (Washington, DC: IFPRI, 2022); M.F. Alvi, P. Barooah, S. Gupta, R.S. Meinzen-Dick, and C. Ringler, "Hello, Can You Hear Me? Impact of Speakerphones on Phone Survey Responses," IFPRI Discussion Paper 2123, IFPRI, Washington, DC, 2022.

52 L. Stark, I. Seff, and C. Reis, "Gender-Based Violence against Adolescent Girls in Humanitarian Settings: A Review of the Evidence," *Lancet Child & Adolescent Health* 5, 3 (2021): 210-222.

53 A. Peterman and M. Ranganathan, "How is Economic Security Linked to Violence against Women and Children? New Insights from the Sexual Violence Research Initiative Forum 2022," *IFPRI Blog*, Nov. 11, 2022.

54 S. Roy, A. Peterman, and L. Billings, "Making Cash Transfers More Effective in Reducing Intimate Partner Violence," *IFPRI Blog*, June 24, 2022.

55 A.M. Thurston, H. Stöckl, and M. Ranganathan, "Natural Hazards, Disasters and Violence against Women and Girls: A Global Mixed-Methods Systematic Review," *BMJ Global Health* 6, 4 (2021): e004377; L. Stark, I. Seff, and C. Reis, "Gender-Based Violence against Adolescent Girls in Humanitarian Settings: A Review of the Evidence," *Lancet Child & Adolescent Health* 5, 3 (2021): 210-222; K.R. van Daalen, S.S. Kallesøe,

F. Davey, et al., "Extreme Events and Gender-Based Violence: A Mixed-Methods Systematic Review," *Lancet Planetary Health* 6,6 (2022): E504–E523.

56 J. Krause, W. Krause, and P. Bränfors, "Women's Participation in Peace Negotiations and the Durability of Peace," *International Interactions* 44, 6 (2018): 985–1016.

57 UN Women, "Women's Participation and a Better Understanding of the Political," in *Preventing Conflict, Transforming Justice, Securing the Peace: A Global Study on the Implementation of United Nations Security Council Resolution 1325*, Ch. 3 (2015).

58 F. Kuwonu, "Women: Liberia's Guardians of Peace," *Africa Renewal* (April–July 2018).

59 Ibid.

60 A.M. Thurston, H. Stöckl, and M. Ranganathan, "Natural Hazards, Disasters and Violence against Women and Girls: A Global Mixed-Methods Systematic Review," *BMJ Global Health* 6, 4 (2021): e004377.

61 ALNAP, *The State of the Humanitarian System* (London: 2022).

62 M. Bizzarri, M.C. Vinas, H. Khogali, and T. Jantzi, *Inter-Agency Humanitarian Evaluation on Gender Equality and the Empowerment of Women and Girls* (New York: UN OCHA, 2020).

63 E. Bryan, C. Ringler, and N. Lefore, "To Ease the World Food Crisis, Focus Resources on Women and Girls," *Nature* 609 (2022): 28–31.

64 E. Bryan, C. Ringler, and N. Lefore, "To Ease the World Food Crisis, Focus Resources on Women and Girls," *Nature* 609 (2022): 28–31; C. Ragasa and I. Lambrecht, "COVID-19 and the Food System: Setback or Opportunity for Gender Equality?" *Food Security* 12 (2020): 877–880.

65 UN Women, "How Six Grass-Roots Women's Organizations Are Making Sure That No One Is Left Behind in COVID-19 Response," *UN Women News*, March 10, 2021.

BOX 1

1 E. Neumayer and T. Plümper, "The Gendered Nature of Natural Disasters: The Impact of Catastrophic Events on the Gender Gap in Life Expectancy, 1981–2002," *Annals of the Association of American Geographers* 97, 3 (2008): 551–556.

2 P.G.D. Chakrabarti and A. Walia, "Toolkit for Mainstreaming Gender in Emergency Response," in *Women, Gender and Disaster: Global Issues and Initiatives*, eds. E. Enarson and P.G.D. Chakrabarti, 337–360 (New Delhi: SAGE Publications, 2009).

3 K. Ikeda, "Gender Differences in Human Loss and Vulnerability in Natural Disasters: A Case Study from Bangladesh," *Indian Journal of Gender Studies* 2, 2 (1995): 171–193.

4 A.M. Thurston, H. Stöckl, and M. Ranganathan, "Natural Hazards, Disasters and Violence against Women and Girls: A Global Mixed-Methods Systematic Review," *BMJ Global Health* 6, 4 (2021): e004377; K. Ikeda, "Gender Differences in Human Loss and Vulnerability in Natural Disasters: A Case Study from Bangladesh," *Indian Journal of Gender Studies* 2, 2 (1995): 171–193.

5 K. Ikeda, "Gender Differences in Human Loss and Vulnerability in Natural Disasters: A Case Study from Bangladesh," *Indian Journal of Gender Studies* 2, 2 (1995): 171–193; J. D'Cunha, "Engendering Disaster Preparedness and Management," *Asian Disaster Management News* 3, 3 (1997): 2–5.

6 C. Sturridge, Z. Ullah, and M.T. Amjad, "Why Some Women in Pakistan Are Being Left Behind in the Floods," *ODI Blog*, Oct. 18, 2022; Agence France Press, "'A Matter of Honour': Women Forced to Stay in Flooded Pakistan Village," Sept. 8, 2022.

BOX 2

1 UNICEF, *Severe Wasting: An Overlooked Child Survival Emergency*, UNICEF Child Alert (May 2022).

2 Ibid.

3 USA for UNHCR, "Famine Explained: Definition, Causes and Facts," Sept. 16, 2022.

4 UNICEF, *Severe Wasting: An Overlooked Child Survival Emergency*, UNICEF Child Alert (May 2022).

CHAPTER 7

1 J.M. Rubenstein, "Migration," in *The Cultural Landscape: An Introduction to Human Geography*, 12th edition, Ch. 3 (London: Pearson, 2017); World Bank, *Moving for Prosperity: Global Migration and Labor Markets*, Policy Research Report (Washington, DC: 2018).

2 M.A. Clemens, C.E. Montenegro, and L. Pritchett, "The Place Premium: Bounding the Price Equivalent of Migration Barriers," *Review of Economics and Statistics* 101, 2 (2019): 201–213; M.A. Clemens and L. Pritchett, "The New Economic Case for Migration Restrictions: An Assessment," *Journal of Development Economics* 138 (2019): 153–164.

3 F. Kondylis and V. Mueller, "Economic Consequences of Conflict and Environmental Displacement," in *International Handbook on Migration and Economic Development*, ed. R.E. Lucas, Ch. 14, 388–424 (Cheltenham, UK, and Northampton, MA: Edward Elgar, 2014).

4 A. Josephson and G.E. Shively, "Unanticipated Events, Perceptions, and Household Labor Allocation in Zimbabwe," *World Development* 141 (2021): 105377; E. Piguet, A. Pecoud, and P. de Guchtneire, "Migration and Climate Change: An Overview," *Refugee Survey Quarterly* 30, 3 (2011): 1–23.

5 According to the 1951 Refugee Convention, a refugee is defined as "someone who is unable or unwilling to return to their country of origin owing to a well-founded fear of being persecuted for reasons of race, religion, nationality, membership of a particular social group, or political opinion."

6 J. Barchfield, "Pandemic Deepens Hunger for Displaced People the World Over," *UNHCR News & Stories*, March 31, 2021; USAID (United States Agency for International Development), *U.S. Government Global Food Security Strategy: Fiscal Year 2022–2026* (Washington, DC: 2021).

7 IOM (International Organization for Migration), *Life Amidst a Pandemic: Hunger, Migration and Displacement in the East and Horn of Africa* (Grand-Saconnex, Switzerland: 2021).

8 A. Steele, "Seeking Safety: Avoiding Displacement and Choosing Destinations in Civil Wars," *Journal of Peace Research* 46, 3 (2009): 419–429; M. Revkin, "To Stay or to Leave? Displacement Decisions during Rebel Governance," Working Paper, Yale University, New Haven, CT, 2019.

9 S. Schmeidl, "Exploring the Causes of Forced Migration: A Pooled Time-Series Analysis, 1971–1990," *Social Science Quarterly* 78, 2 (1997): 284–308; J. Schon, "Motivation and Opportunity for Conflict-Induced Migration: An Analysis of Syrian Migration Timing," *Journal of Peace Research* 56, 1 (2019): 12–27.

10 S. Engel and A.M. Ibáñez, "Displacement Due to Violence in Colombia: A Household-Level Analysis," *Economic Development and Cultural Change* 55, 2 (2007): 335–365; P. Adhikari, "Conflict-Induced Displacement, Understanding the Causes of Flight," *American Journal of Political Science* 57, 1 (2013): 82–89.

11 J. Schon, "Motivation and Opportunity for Conflict-Induced Migration: An Analysis of Syrian Migration Timing," *Journal of Peace Research* 56, 1 (2019): 12-27.

12 IOM, "Migration in the World," accessed January 2023.

13 FAO (Food and Agriculture Organization of the United Nations), *Migration, Agriculture, and Rural Development: Addressing the Root Causes of Migration and Harnessing Its Potential for Development* (Rome: 2016); IOM, *World Migration Report 2020* (Geneva: 2020).

14 IOM, "Key Migration Terms," accessed January 2023.

15 A. Kraler and D. Reichel, "Measuring Irregular Migration and Population Flows: What Available Data Can Tell," *International Migration* 49, 5 (2011): 97-128.

16 More generally, the pandemic was associated with a temporary slow-down in international migration, with an estimated decrease of nearly 2 million international migrants (equivalent to 27 percent) between March and July 2020; UN DESA, "Growth of International Migration Slowed by 27%, or 2 Million Migrants, due to COVID-19, says UN," accessed February 2023.

17 U.S. Customs and Border Protection website, accessed Feb. 2023.

18 F. Ceballos and M.A. Hernandez, "The Migration Propensity Index: An Application to Guatemala," IFPRI Discussion Paper 1953, IFPRI, Washington, DC, 2020.

19 ACCS (Africa Center for Strategic Studies), "African Migration Trends to Watch in 2022," Dec. 17, 2021; Pew Research Center, "Middle East's Migrant Population More than Doubles since 2005," Oct. 18, 2016.

20 S. Bermeo, D. Lebland, and G. Nagle Alverio, "Root Causes of Migration from Guatemala: Analysis of Subnational Trends," Policy Brief (Durham: Duke Center for International Development, 2022); D. Cohn, J.S. Passel, and A. Gonzalez-Barrera, *Rise in U.S. Immigrants From El Salvador, Guatemala and Honduras Outpaces Growth from Elsewhere* (Washington, DC: Pew Research Center, 2017); CRS (Congressional Research Service), *Recent Migration to the United States from Central America: Frequently Asked Questions* (Washington, DC: 2019); CRS, "Central American Migration: Root Causes and U.S. Policy," in Focus IFII151 (Washington, DC: 2019); National Immigration Forum, *Push or Pull Factors: What Drives Central American Migrants to the U.S.?* (Washington, DC: 2019).

21 UNHCR (United Nations High Commissioner for Refugees), "Operational Data Portal: Ukrainian Refugee Situation," accessed January 2023. https://data.unhcr.org/en/situations/ukraine

22 IOM, "Migration in the World," accessed January 2023.

23 UNGA (United Nations General Assembly), "The New York Declaration on Refugees and Migrants," article I.1 (Washington, DC: 2016).

24 V. Mueller, C. Gray, and K. Kosec, "Heat Stress Increases Long-Term Human Migration in Rural Pakistan," *Nature Climate Change* 4 (2014): 182-185; J. Milán-García, J.L. Caparrós-Martínez, N. Rueda-López, and J.D.P. Valenciano, "Climate Change-Induced Migration: A Bibliometric Review," *Global Health* 17 (2021): 74.

25 M. Grecequet, J. DeWaard, J.J. Hellmann, and G.J. Abel, "Climate Vulnerability and Human Migration in Global Perspective," *Sustainability* 9, 5 (2017): 720.

26 K. Jessoe, D.T. Manning, and J.E. Taylor, "Climate Change and Labour Allocation in Rural Mexico: Evidence from Annual Fluctuations in Weather," *Economic Journal* 128, 608 (2018): 230-261.

27 P. Mahajan and D. Yang, "Taken by Storm: Hurricanes, Migrant Networks, and US Immigration," *American Economic Journal: Applied Economics* 12, 2 (2020): 250-277.

28 P. Läderach, J. Ramirez-Villegas, G. Caroli, C. Sadoff, and G. Pacillo, "Climate Finance and Peace: Tackling the Climate and Humanitarian Crisis," *Lancet Planetary Health* 5, 12 (2021): e856-e858.

29 OHCHR (Office of the UN High Commissioner for Human Rights), "Climate Chang Exacerbates Violence against Women and Girls," *OHCHR Stories*, July 12, 2022.

30 CGIAR Research Program on Policies, Institutions, and Markets, *Understanding Characteristics, Causes, and Consequences of Migration: Contributions from the CGIAR Research Program on Policies, Institutions and Markets*, PIM Synthesis Brief September 2021 (Washington, DC: IFPRI, 2021).

31 K. Kosec, H. Ghebru, B. Holtemeyer, V. Mueller, and E. Schmidt, "The Effect of Land Access on Youth Employment and Migration Decisions: Evidence from Rural Ethiopia," *American Journal of Agricultural Economics* 100, 3 (2018): 931-954; M. Rufai, A. Ogunniyi, K.K. Salman, M. Oyeyemi, and M. Salawum "Migration, Labor Mobility and Household Poverty in Nigeria: A Gender Analysis," *Economies* 7, 4 (2019): 101.

32 K. Kosec, J. Song, H. Zhao, and B. Holtemeyer, "The Gendered Impacts of Income Fluctuations on Household Departure, Labor Supply, and Human Capital Decisions: Evidence from Kyrgyzstan," IFPRI Discussion Paper 2040, IFPRI, Washington, DC, 2021.

33 A. De Brauw and J. Giles, "Migrant Opportunity and the Educational Attainment of Youth in Rural China," *Journal of Human Resources* 52, 1 (2017): 272-311; H. Ghebru, M. Amare, G. Mavrotas, and A. Ogunniyi, "Role of Land Access in Youth Migration and Youth Employment Decisions: Empirical Evidence from Rural Nigeria," NSSP (Nigeria Strategy Support Program) Working Paper 58, IFPRI, Washington, DC, 2018; K. Kosec, H. Ghebru, B. Holtemeyer, V. Mueller, and E. Schmidt, "The Effect of Land Access on Youth Employment and Migration Decisions: Evidence from Rural Ethiopia," *American Journal of Agricultural Economics* 100, 3 (2018): 931-954; K. Kosec, J. Song, H. Zhao, and B. Holtemeyer, "The Gendered Impacts of Income Fluctuations on Household Departure, Labor Supply, and Human Capital Decisions: Evidence from Kyrgyzstan," IFPRI Discussion Paper 2040, IFPRI, Washington, DC, 2021.

34 J. Baez, G. Caruso, V. Mueller, and C. Niu, "Droughts Augment Youth Migration in Northern Latin America and the Caribbean," *Climatic Change* 140 (2017): 423-435; K. Hirvonen, "Temperature Changes, Household Consumption, and Internal Migration: Evidence from Tanzania," *American Journal of Agricultural Economics* 98, 4 (2016): 1230-1249; B. Thiede, C. Gray, and V. Mueller, "Climate Variability and Inter-provincial Migration in South America, 1970-2011," *Global Environmental Change* 41 (2016): 228-240.

35 C. Gray, D. Hopping, and V. Mueller, "The Changing Climate-Migration Relationship in China, 1989-2011," *Climatic Change* 160 (2020): 103-122.

36 K. Ambler, "Migration and Remittances in Central America: New Evidence and Pathways for Future Research," LAC Working Paper 4, IFPRI, Washington, DC, 2019; E. Masset, A. Gelli, J-P. Tranchant, and A.S. Diallo, "Impact of Conflict on Agriculture in Mali," IFPRI Discussion Paper 1843, IFPRI, Washington, DC, 2019.

37 J.F. Hoddinott and T.J. Mekasha, "Social Protection, Household Size and Its Determinants: Evidence from Ethiopia," ESSP (Ethiopia Strategy Support Program) Working Paper 107, IFPRI and Ethiopia Development Research Institute (EDRI), Washington, DC and Addis Ababa, 2017; V. Mueller, C. Gray, S. Handa, and D. Seidenfeld, "Do Social Protection Programs Foster Short-Term and Long-Term Migration Adaptation Strategies?" *Environment and Development Economics* 25, 2 (2020): 135-158; M. Hidrobo, V. Mueller, and S. Roy, "Cash Transfers, Migration, and Gender Norms," *American Journal of Agricultural Economics* 104, 2 (2021): 550-568.

38 J. Chen, K. Kosec, and V. Mueller, "Moving to Despair? Migration and Well-Being in Pakistan," *World Development* 113 (2019): 186-203.

39 A. De Brauw, V. Mueller, and T. Woldehanna, "Does Internal Migration Improve Overall Well-Being in Ethiopia?" *Journal of African Economies* 27, 3 (2018): 347-365; J. De Weerdt and K. Hirvonen, "Risk Sharing and Internal Migration," *Economic Development and Cultural Change* 65, 1 (2016): 63-86; FAO, *The State of Food and Agriculture 2018: Migration, Agriculture and Rural Development* (Rome: 2018).

40 J. Chen, K. Kosec, and V. Mueller, "Moving to Despair? Migration and Well-Being in Pakistan," *World Development* 113 (2019): 186-203; A. Fabry, B. van den Broeck, and M. Maertens, "Decent Work in Global Food Value Chains: Evidence from Senegal," Paper presented at the 3rd Annual IZA/World Bank/NJD/ UNU-WIDER Jobs and Development Conference: Better Jobs for Development, 2020; K. Hirvonen and H.B. Lilleør, "Going Back Home: Internal Return Migration in Rural Tanzania," Sussex Centre for Migration Research (SCMR) Working Paper 77, University of Sussex, Brighton, UK, 2014; A. Mattoo, I.C. Neagu, and C. Özden, "Brain Waste: Educated Immigrants in the US Labor Market," Policy Research Paper 3581, World Bank, Washington, DC, 2005.

41 R. Donnelly and J.M. Hagan, "The Dangerous Journey: Migrant Smuggling from Mexico and Central America, Asia, and the Caribbean," in *Hidden Lives and Human Rights in the United States: Understanding the Controversies and Tragedies of Undocumented Immigration*, ed. L.A. Lorentzen, Ch. 4, 71-106 (Santa Barbara: Praeger, 2014); A. Isacson, M. Meyer, and G. Morales, *Mexico's Other Border: Security, Migration, and the Humanitarian Crisis at the Line with Central America* (Washington, DC: Washington Office on Latin America, 2014).

42 J. Black and Z. Sigman, "50,000 Lives Lost during Migration: Analysis of Missing Migrants Project data 2014-2022,"(IOM GMDAC: 2022).

43 R.H.J. Adams and A. Cuecuecha, "Remittances, Household Expenditure and Investment in Guatemala, *World Development* 38, 11 (2010): 1626-1641; K. Ambler, D. Aycinena, and D. Yang, "Channeling Remittances to Education: A Field Experiment among Migrants from El Salvador," *American Economic Journal: Applied Economics* 7, 2 (2015): 207-232; A. De Brauw, "Migration Out of Rural Areas and Implications for Rural Livelihoods," *Annual Review of Resource Economics* 11, 1 (2019): 461-481; A. De Brauw and J. Giles, "Migrant Labor Markets and the Welfare of Rural Households in the Developing World: Evidence from China," *World Bank Economic Review* 32, 1 (2018): 1-18; K. Mishra, O. Kondratjeva, and G.E. Shively, "Do Remittances Reshape Household Expenditures? Evidence from Nepal," *World Development* 157 (2022): 105926; R. Mu, and A. de Brauw, "Migration and Young Child Nutrition: Evidence from Rural China," *Journal of Population Economics* 28, 3 (2015): 631-657; D. Yang and C. Martinez, "Remittances and Poverty in Migrants' Home Areas: Evidence from the Philippines," in *International Migration, Remittances and the Brain Drain*, eds. C. Ozden and M. Schiff, Ch. 3, 151-199. (Washington, DC: World Bank, 2006).

44 R. Maldonado and J. Harris, "Las Remesas a Latinoamérica y el Caribe en 2021: El Esfuerzo de los Migrantes Durante la Crisis del COVID-19," Nota Técnica N° IDB-TN-2493, Inter-American Development Bank (IDB), Washington, DC, 2022.

45 D. Ratha, E.J. Kim, S. Plaza, E.J. Riordan, and V. Chandra, "A War in a Pandemic: Implications of the Ukraine Crisis and COVID-19 on Global Governance of Migration and Remittance Flows," Migration and Development Brief 36 (Washington, DC: World Bank Group, 2022).

46 M.J. Filipski, H.L. Lee, A. Hein, and U. Nischan, "Emigration and Rising Wages in Myanmar: Evidence from Mon State," *Journal of Development Studies* 56, 5 (2020): 946-963.

47 N. Kawarazuka, T.M. Duong, and E. Simelton, "Gender, Labor Migration and Changes in Small-Scale Farming on Vietnam's North-Central Coast," *Critical Asian Studies* 52, 4 (2020): 550-564; V. Mueller, C. Doss, and A. Quisumbing, "Youth Migration and Labour Constraints in African Agrarian Households," *Journal of Development Studies* 54, 5 (2018): 875-894.

48 J.N. Baada and D. Najjar, "A Review of the Effects of Migration on the Feminization of Agrarian Dryland Economies," *Journal of Gender, Agriculture and Food Security* 5, 2 (2020): 1-12; C. Doss, A. Qaisrani, K. Kosec, V. Slavchevska, A. Galiè, and N. Kawarazuka, "From the "Feminization of Agriculture" to Gender Equality," in *Advancing Gender Equality through Agricultural and Environmental Research: Past, Present, and Future*, eds. R. Pyburn, and A. van Eerdewijk, Ch. 8, 297-327 (Washington, DC: IFPRI, 2021).

49 M. Beine, F. Docquier, and H. Rapoport, "Brain Drain and Human Capital Formation in Developing Countries: Winners and Losers," *Economic Journal* 118, 528 (2008): 631-652; F. Docquier and A. Marfouk, "International Migration by Education Attainment, 1990-2000," in *International Migration, Remittances and the Brain Drain*, eds. C. Ozden and M. Schiff, Ch. 5, 151-199 (Washington DC: World Bank, 2006).

50 S. Berger, "Brain Drain, Brain Gain and Its Net Effect," KNOMAD Paper No. 46, World Bank, Washington, DC, 2022.

51 J-F. Maystadt, K. Hirvonen, A. Mabiso, and J. Vandercasteelen, "Impacts of Hosting Forced Migrants in Poor Countries," *Annual Review of Resource Economics* 11 (2019): 439-459; I. Ruiz and C. Vargas-Silva, "The Economics of Forced Migration," *Journal of Development Studies* 49, 6 (2013): 772-784; P. Verme and K. Schuettler, "The Impact of Forced Displacement on Host Communities: A Review of the Empirical Literature in Economics," Policy Research Working Paper 8727, World Bank, Washington, DC, 2021.

52 J-F. Maystadt and P. Verwimp, "Winners and Losers among a Refugee-Hosting Population," *Economic Development and Cultural Change* 62, 4 (2014): 769-809; J.E. Taylor, M.J. Filipski, M. Alloush, A. Gupta, R.I.R. Valdes, and E. Gonzalez-Estrada, "Economic Impact of Refugees," *Proceedings of the National Academy of Sciences* 113, 27 (2016): 7449-7453; J-F. Maystadt and G. Duranton, "The Development Push of Refugees: Evidence from Tanzania," *Journal of Economic Geography* 19, 2 (2019): 299-334.

53 J.E. Taylor, M.J. Filipski, M. Alloush, A. Gupta, R.I.R. Valdes, and E. Gonzalez-Estrada, "Economic Impact of Refugees," *PNAS* 113, 27 (2016): 7449-7453.

54 J-F. Maystadt, K. Hirvonen, A. Mabiso, and J. Vandercasteelen, "Impacts of Hosting Forced Migrants in Poor Countries," *Annual Review of Resource Economics* 11 (2019): 439-459.

55 I. Ruiz and C. Vargas-Silva, "The Economics of Forced Migration," *Journal of Development Studies* 49, 6 (2013): 772-784.

56 E. Aksu, R. Erzan, and M. Güray Kırdar, "The Impact of Mass Migration of Syrians on the Turkish Labor Market," IZA Discussion Papers 12050, Institute of Labor Economics (IZA), Bonn, 2022; G. Caruso, C.G. Canon, and V. Mueller, "Spillover Effects of the Venezuelan Crisis: Migration Impacts in Colombia," *Oxford Economic Papers* 73, 2 (2021): 771-795; B. Malaeb and J. Wahba, "Migration Dynamics during the Refugee Influx in Jordan," in *The Jordanian Labor Market: Between Fragility and Resilience*, eds. C. Krafft R. and Assaad, Ch. 5, 125-140 (Oxford: Oxford University Press, 2019). The peak of the Syrian civil war in 2015 and the resulting large inflows of Syrian refugees into neighboring countries created new momentum that drew the attention of policymakers and researchers to forced migration and its effects on hosting communities, including effects beyond labor markets, and marked a shift in the broader migration literature.

57 Others have also studied the impact of refugees on other outcomes, such as business openings (Y.E. Akgündüz, M. Van den berg, and W.

Hassink, "The Impact of the Syrian Refugee Crisis on Firm Entry and Performance in Turkey," *World Bank Economic Review* 32, 1 (2018): 19–40; O. Altındağ, O. Bakış, and S.V. Rozo, "Blessing or Burden? Impacts of Refugees on Businesses and the Informal Economy," *Journal of Development Economics* 146 (2020): 102490), the environment (C.G. Aksoy and S. Tumen, "Local Governance Quality and the Environmental Cost of Forced Migration," *Journal of Development Economics* 149 (2021): 102603; J-F. Maystadt, V. Mueller, J. Van Den Hoek, and S. van Weezel, "Vegetation Changes Attributable to Refugees in Africa Coincide with Agricultural Deforestation," *Environmental Research Letters* 15, 4 (2020): 044008), and the propagation of diseases (A. Ibáñez, S. Rozo, and M. Urbina, "Forced Migration and the Spread of Infectious Diseases," *Journal of Health Economics* 79 (2020): 102491).

58 A. Kayaoglu, "Do Refugees Cause Crime?" *World Development* 154 (2022): 105858; B.G. Knight and A. Tribin, "Immigration and Violent Crime: Evidence from the Colombia-Venezuela Border," NBER Working Paper 27620, National Bureau of Economic Research, Cambridge, MA, 2020.

59 Y. Zhou and A. Shaver, "Reexamining the Effect of Refugees on Civil Conflict: A Global Subnational Analysis," *American Political Science Review* 115, 4 (2021): 1175–1196.

60 M. Beine, L. Bertinelli, R. Cömertpay, A. Litina, and J-F. Maystadt, "A Gravity Analysis of Refugee Mobility Using Mobile Phone Data," *Journal of Development Economics* 150 (2021): 102618; A. Milliff and F. Christia, "Who Flees Conflict? A Big-Data Approach to the Determinants of Forced Migration," Working Paper, Stanford University and The Global Diversity Lab, Redwood City, CA, and Cambridge, MA, 2021.

61 C. MacPherson and O. Sterck, "Empowering Refugees through Cash and Agriculture: A Regression Discontinuity Design," HiCN Working Papers 320, Households in Conflict Network, Berlin, 2021; M. Alloush, J.E. Taylor, A. Gupta, R.I. Rojas, and E. Gonzalez-Estrada, "Economic Life in Refugee Camps," *World Development* 95 (2017): 334–347; M. Hidrobo, J. Hoddinott, A. Peterman, A. Margolies, and V. Moreira, "Cash, Food, or Vouchers? Evidence from a Randomized Experiment in Northern Ecuador," *Journal of Development Economics* 107 (2014): 144–156.

62 J. Gazeaud, E. Mvukiyehe, and O. Sterck, "Cash Transfers and Migration: Theory and Evidence from a Randomized Controlled Trial," *Review of Economics and Statistics* 105 (2023): 143–157.

63 P. Läderach, J. Ramirez-Villegas, G. Caroli, C. Sadoff, and G. Pacillo, "Climate Finance and Peace: Tackling the Climate and Humanitarian Crisis," *Lancet Planetary Health* 5, 12 (2021): e856–e858.

64 The project supported by the German government's International Climate Initiative is known as "Implementing Sustainable Land Use Systems to Contribute to Forest Conservation, Climate Protection (REDD+) and the Peace-Building Process in Colombia."

65 A. Betts, M.F. Stierna, N. Omata, and O. Sterck, "Social Cohesion and Refugee-Host Interactions: Evidence from East Africa," Policy Research Working Paper 9917, World Bank, Washington, DC, 2022; C. Blair, G. Grossman, and J.M. Weinstein, "Liberal Displacement Policies Attract Forced Migrants in the Global South," *American Political Science Review* 116, 1 (2021): 351–358; M.M. Kadigo, N.O. Diallo, and J-F. Maystadt, "How to Cope with a Refugee Shock?" Policy Research Working Paper 9950, World Bank, Washington, DC, 2022; A. Ibáñez, A. Moya, M.A. Ortega, S.V. Rozo, and M.J. Urbina, "Life Out of the Shadows: Impacts of Amnesties in the Lives of Refugees," Policy Research Working Paper 9928, World Bank, Washington, DC, 2022.

66 C. Schneiderheinze and M. Lücke, "Socio-economic Impacts of Refugees on Host Communities in Developing Countries," PEGNet Policy Studies03/2020, Kiel Institute for the World Economy (IfW), Poverty Reduction, Equity and Growth Network (PEGNet), Kiel, Germany, 2020.

67 M. Beine, F. Docquier, and H. Rapoport, "Brain Drain and Human Capital Formation in Developing Countries: Winners and Losers," *Economic Journal* 118, 528 (2008): 631–652.

68 J. Nielsen Arendt, C. Dustmann, and H. Ku, "Refugee Migration and the Labor Market: Lessons from 40 Years of Post-Arrival Policies in Denmark," CESifo Discussion Paper 17184, Center for Economic Studies, Munich, 2022.

69 M. Lehmann and D. Masterson, "Does Aid Reduce Anti-Refugee Violence? Evidence from Syrian Refugees in Lebanon," *American Political Science Review* 114, 4 (2020): 1335–1342.

70 O. Altindağ and N. Kaushal, "Do Refugees Impact Voting Behavior in the Host Country? Evidence from Syrian Refugee Inflows in Turkey," IZA Discussion Papers 10849, Institute of Labor Economics (IZA), Bonn, 2017; S. Rozo and J.F. Vargas, "Brothers or Invaders? How Crisis-Driven Migrants Shape Voting Behavior," LACEA Working Paper Series No. 0014, Latin American and Caribbean Economic Association, Bogotá, 2018; Y. Zhou and G. Grossman, "When Refugee Presence Increases Incumbent Support through Development: Evidence from Uganda," OSF Preprints, available online from Feb. 2023.

71 Y. Zhou, G. Grossman, and S. Ge, "Inclusive Refugee-Hosting in Uganda Improves Local Development and Prevents Public Backlash," Policy Research Working Paper 9981, World Bank, Washington, DC, 2022.

REGIONAL DEVELOPMENTS

AFRICA

1 K. Reid, "Africa Hunger, Famine: Facts, FAQs, and How to Help," World Vision, Feb. 22, 2022.

2 FSIN (Food Security Information Network) and GNAFC (Global Network Against Food Crises), *2022 Global Report on Food Crises* (Rome: 2022).

3 IPC (Integrated Food Security Phase Classification), "South Sudan: IPC Acute Food Insecurity & Malnutrition Snapshot, Acute Food Insecurity: October 2022–July 2023; Acute Malnutrition July 2022–June 2023" (2022).

4 W. Drammeh, N.A. Hamid, and A.J. Rohana, "Determinants of Household Food Insecurity and Its Association with Child Malnutrition in sub-Saharan Africa: A Review of the Literature," *Current Research in Nutrition and Food Science* 7, 3 (2019); A.A. Misselhorn, "What Drives Food Insecurity in Southern Africa? A Meta-Analysis of Household Economy Studies," *Global Environmental Change* 15, 1 (2005): 33–43.

5 J. Wan, et al., "Biology, Invasion and Management of the Agricultural Invader: Fall Armyworm, *Spodoptera frugiperda* (Lepidoptera: Noctuidae)," *Journal of Integrative Agriculture* 20, 3 (2021): 646–663.

6 Y. Dong, L. Zhao, and W. Huang, *Monitoring of Desert Locust in Africa and Asia* (Beijing: Springer Nature, 2023).

7 V. Bjornlund, H. Bjornlund, and A. van Rooyen, "Why Food Insecurity Persists in sub-Saharan Africa: A Review of Existing Evidence," *Food Security* 14 (2022): 845–864; S. Benin, S. Wood, and A. Nin-Pratt, "Introduction," in *Agricultural Productivity in Africa: Trends, Patterns, and Determinants*, S. Benin, ed. (Washington, DC: IFPRI, 2016).

8 For further discussion of the COVID-19 pandemic and effects in Africa, see IFPRI, *2021 Global Food Policy Report: Transforming Food Systems after COVID-19* (Washington, DC: 2021).

9 J. Sacko and I. Mayaki, "How the Russia-Ukraine Conflict Impacts Africa: An Opportunity to Build Resilient, Inclusive Food Systems in Africa," *Africa Renewal* (April 21, 2022); V. Bilger and N.B. Brahim,

Could Disruptions to Eastern European Wheat Spur Displacement in Africa?, ICMPD blogpost, International Centre for Migration Policy Development, May 10, 2022.

10 ACLED, ACLED data dashboard, accessed December 6, 2022. https://acleddata.com/dashboard

11 FAO (Food and Agriculture Organization of the United Nations), IFAD (International Fund for Agricultural Development), UNICEF, WFP (World Food Programme), and WHO (World Health Organization), *The State of Food Security and Nutrition in the World 2022: Repurposing Food and Agricultural Policies to Make Healthy Diets More Affordable* (Rome: FAO, 2022).

12 FAO, "New FAO Data Highlights Gender Gap in Food Insecurity across Regions," Sept. 25, 2017.

13 Oxfam, *The Cost of Hunger* (Johannesburg: 2014).

14 A.B. Kabia, "Women in Sierra Leone's Labour Market," *China Daily*, Feb. 9, 2016.

15 D. Casale and D. Shepherd, "The Gendered Effects of the Covid-19 Crisis in South Africa: Evidence from NIDS-CRAM Waves 1-5," *Development Southern Africa* (2022): 1-20.

16 IRC (International Rescue Committee), "Women and Girls in DRC Facing an Increased Risk of Violence and Higher Exposure to Ebola since the Start of the Outbreak," press release, March 2019.

17 C. Wenham, J. Smith, S.E. Davies, et al., "Women Are Most Affected by Pandemics: Lessons from Past Outbreaks," comment, *Nature* (July 8, 2020).

18 M. Altman, T. Hart, and P. Jacobs, *Food Security in South Africa* (Pretoria: Human Sciences Research Council, 2009).

19 T. Bundervoet, P. Verwimp, and R. Akresh, "Health and Civil War in Rural Burundi," *Journal of Human Resources* 44, 2 (2009): 536-563.

20 C. Minoiu and O.N. Shemyakina, "Armed Conflict, Household Victimization, and Child Health in Côte d'Ivoire," *Journal of Development Economics* 108 (2014): 237-255.

21 R. Akresh, L. Lucchetti, and H. Thirumurthy, "Wars and Child Health: Evidence from the Eritrean-Ethiopian Conflict," *Journal of Development Economics* 99, 2 (2012): 330-340.

22 G. Dunn, "The Impact of the Boko Haram Insurgency in Northeast Nigeria on Childhood Wasting: A Double-Difference Study," *Conflict and Health* 12, 1 (2012): 1-12; E. Howell, T. Waidmann, N. Holla, N. Birdsall, and K. Jiang, "The Impact of Civil Conflict on Child Malnutrition and Mortality, Nigeria, 2002-2013," CGD Working Paper No. 494, Center for Global Development, Washington, DC, 2018.

23 R. Akresh, P. Verwimp, and T. Bundervoet, "Civil War, Crop Failure, and Child Stunting in Rwanda," *Economic Development and Cultural Change* 59, 4 (2011): 777-810.

24 D.K. Kinyoki, G.M. Moloney, O.A. Uthman, N.B. Kandala, E.O. Odundo, A.M. Noor, and J.A. Berkley, "Conflict in Somalia: Impact on Child Undernutrition," *BMJ Global Health* 2, 2 (2017): e000262.

25 D. Backer and T. Billing, "Validating Famine Early Warning Systems Network Projections of Food Security in Africa, 2009-2020," *Global Food Security* 29 (2021): 100510; P.K. Krishnamurthy, R.J. Choularton, and P. Kareiva, "Dealing with Uncertainty in Famine Predictions: How Complex Events Affect Food Security Early Warning Skill in the Greater Horn of Africa," *Global Food Security* 26 (2020): 100374; D. Maxwell, *Famine Early Warning and Information Systems in Conflict Settings: Challenges for Humanitarian Metrics and Response* (Boston: Feinstein International Center, Tufts University, 2019).

26 See Africa CDC (Centres for Disease Control and Prevention), https://africacdc.org

27 I.S. Fall, S. Rajatonirina, A.A. Yahaya, et al., "Integrated Disease Surveillance and Response (IDSR) Strategy: Current Status, Challenges and Perspectives for the Future in Africa," *BMJ Global Health* 4 (2019): e001427; C.M. Wolfe, E.L. Hamblion, E.K. Dzotsi, et al., "Systematic Review of Integrated Disease Surveillance and Response (IDSR) Implementation in the African Region," *PLoS ONE* 16, 2 (2021): e0245457.

28 CGIAR Research Program on Policies, Institutions, and Markets (PIM), *Understanding Characteristics, Causes, and Consequences of Migration: Contributions from the CGIAR Research Program on Policies, Institutions, and Markets*, PIM Synthesis Brief (Washington, DC: IFPRI, 2021).

29 D. Hilhorst, "Classical Humanitarianism and Resilience Humanitarianism: Making Sense of Two Brands of Humanitarian Action," *Journal of International Humanitarian Action* 3, 15 (2018).

30 WFP, *Annual Achievements 2021* (Dakar: WFP, Regional Bureau for Western Africa, 2022); WFP, *Food Assistance for Assets (FFA) for Zero Hunger and Resilient Livelihoods: A Programme Guidance Manual* (Rome: WFP Policy and Programme Division, 2016).

31 S.S. Patel, M.B. Rogers, R. Amlôt, and G.J. Rubin, "What Do We Mean by 'Community Resilience'? A Systematic Literature Review of How It Is Defined in the Literature," *PLOS Currents* 9 (Feb. 1, 2017); M.A. Constas, L. Mattioli, and L. Russo, "What Does Resilience Imply for Development Practice? Tools for More Coherent Programming and Evaluation of Resilience," *Development Policy Review* 39 (2021): 588-603; FAO, *RIMA-II: Resilience Index Measurement and Analysis II* (Rome: 2016).

32 M. Fitzpatrick, K. Spears, J. Ryan, S. Polzin, G. Gottlieb, and D. Maxwell, *Making the Nexus Real: Moving from Theory to Practice* (Boston: Feinstein International Center, Tufts University, 2021); P. Howe, "The Triple Nexus: A Potential Approach to Supporting the Achievement of the Sustainable Development Goals?" *World Development* 124 (2019): 104629; OECD (Organisation for Economic Co-operation and Development), *DAC Recommendation on the Humanitarian-Development-Peace Nexus*, OECD Legal Instruments 5019 (Paris: 2020).

33 M. Fitzpatrick, K. Spears, J. Ryan, S. Polzin, G. Gottlieb, and D. Maxwell, *Making the Nexus Real: Moving from Theory to Practice* (Boston: Feinstein International Center, Tufts University, 2021). See also South Sudan Partnership for Resilience, www.southsudanpfrr.org.

34 See Fighting Food Crises along the Humanitarian-Development-Peace Nexus Coalition, https://www.fightfoodcrises.net/hdp-coalition.

35 M. Fitzpatrick, K. Spears, J. Ryan, S. Polzin, G. Gottlieb, and D. Maxwell, *Making the Nexus Real: Moving from Theory to Practice* (Boston: Feinstein International Center, Tufts University, 2021); P. Howe, "The Triple Nexus: A Potential Approach to Supporting the Achievement of the Sustainable Development Goals?" *World Development* 124 (2019): 104629; UNOCHA (UN Office for the Coordination of Humanitarian Affairs), *Sahel Crisis: Humanitarian Needs and Requirements Overview* (2022).

36 Globally, the share of official development assistance (ODA) allocated for humanitarian assistance increased from less than 5 percent in the 1990s to about 23 percent in 2019. By 2019, 43 countries had experienced periods of high humanitarian assistance lasting longer than 10 years, with an increasing incidence in the past two decades. Of these 43 countries, 15 are in Africa (G. Milante and J. Lylja, *Chronic Crisis Financing? Fifty Years of Humanitarian Aid and Future Prospects*, SIPRI Insights on Peace and Security No. 22/5, April 2022).

37 The financing gap (investment-to-GDP ratio) between 2015 and 2030 to end extreme poverty by 2030, for example, is estimated to be about 66 percent per year (A. Kedir, A. Elhiraika, Z. Chinzara, and D. Sandjong, "Growth and Development Finance Required for Achieving Sustainable

Development Goals (SDGs) in Africa," *African Development Review* 29, S1 (2017): 15–26).

38 FAO, IFAD, UNICEF, WFP and WHO, *The State of Food Security and Nutrition in the World 2022: Repurposing Food and Agricultural Policies to Make Healthy Diets More Affordable* (Rome: FAO, 2022).

BOX 1

1 See Great Green Wall, www.greatgreenwall.org

2 D. Goffner, H. Sinare, and L.J. Gordon, "The Great Green Wall for the Sahara and the Sahel Initiative as an Opportunity to Enhance Resilience in Sahelian Landscapes and Livelihoods," *Regional Environmental Change* 19 (2019): 1417-1428; WFP (World Food Programme), *Greening the Sahel: How WFP's Integrated Resilience Programme Generates Climate Solutions at Scale* (Rome: 2022).

3 M.D. Turner, T. Carney, L. Lawler, J. Reynolds, L. Kelly, M.S. Teague, and L. Brottem, "Environmental Rehabilitation and the Vulnerability of the Poor: The Case of the Great Green Wall," *Land Use Policy* 111 (2021): 105750; A. Mirzabaev, M. Sacande, F. Motlagh, A. Shyrokaya, and A. Martucci, "Economic Efficiency and Targeting of the African Great Green Wall," *Nature Sustainability* 5 (2022): 17–25.

4 M.D. Turner, T. Carney, L. Lawler, J. Reynolds, L. Kelly, M.S. Teague, and L. Brottem, "Environmental Rehabilitation and the Vulnerability of the Poor: The Case of the Great Green Wall," *Land Use Policy* 111 (2021): 105750.

5 A. Mirzabaev, M. Sacande, F. Motlagh, A. Shyrokaya, and A. Martucci, "Economic Efficiency and Targeting of the African Great Green Wall," *Nature Sustainability* 5 (2022): 17–25.

MIDDLE EAST AND NORTH AFRICA

1 World Bank, *Beyond Scarcity: Water Security in the Middle East and North Africa*, MENA Development Report (Washington, DC: 2018).

2 K.A. Abay, C. Breisinger, J. Glauber, S. Kurdi, D. Laborde, and K. Siddig, "The Russia-Ukraine Crisis: Implications for Global and Regional Food Security and Potential Policy Responses," *Global Food Security* 36 (2023): 10675.

3 Ibid.

4 D. Laborde and A. Mamun, "Documentation for Food and Fertilizers Export Restriction Tracker: Tracking Export Policy Responses Affecting Global Food Markets during Crisis," Food and Fertilizer Trade Policy Working Paper 2, IFPRI, Washington, DC, 2022.

5 F. Belhaj, R. Gatti, D. Lederman, E.J. Sergenti, H. Assem, R. Lotfi, and M.E. Mousa, *A New State of Mind: Greater Transparency and Accountability in the Middle East and North Africa* (Washington, DC: World Bank, 2022).

6 K.A. Abay, F. Abdelradi, C. Breisinger, et al., "Egypt: Impacts of the Ukraine and Global Crises on Poverty and Food Security," Global Crisis Country Series Brief 18 (Washington, DC: IFPRI, 2022).

7 J. Glauber and D. Laborde, "How Sanctions on Russia and Belarus Are Impacting Exports of Agricultural Products and Fertilizer," *IFPRI Blog: Issue Post*, November 9, 2022.

8 D.D. Headey and H.H. Alderman, "The Relative Caloric Prices of Healthy and Unhealthy Foods Differ Systematically across Income Levels and Continents," *Journal of Nutrition* 149, 11 (2019): 2020-2033; K. Hirvonen, Y. Bai, D. Headey, W.A. Masters, "Affordability of the EAT-Lancet Reference Diet: A Global Analysis," *Lancet Global Health* 8, 1 (2019): E59-E66.

9 K.A. Abay, C. Breisinger, J. Glauber, S. Kurdi, D. Laborde, and K. Siddig, "The Russia-Ukraine Crisis: Implications for Global and Regional Food Security and Potential Policy Responses," *Global Food Security* 36 (2023): 10675.

10 S. Kurdi, "The Nutritional Benefits of Cash Transfers in Humanitarian Crises: Evidence from Yemen," *World Development* 148 (2021): 105664.

11 K.A. Abay, N. Yonzan, S. Kurdi, and K. Tafere, "Revisiting Poverty Trends and the Role of Social Protection Systems in Africa during the COVID-19 Pandemic (English)," Policy Research Working Paper No. WPS10172, World Bank, Washington, DC, 2022.

12 F. Belhaj, R. Gatti, D. Lederman, E.J. Sergenti, H. Assem, R. Lotfi, and M.E. Mousa, *A New State of Mind: Greater Transparency and Accountability in the Middle East and North Africa* (Washington, DC: World Bank, 2022).

13 U. Gentilini, M.B.A. Almenfi, G. Valleriani, Y. Okamura, E.R. Urteaga, and S. Aziz, "Tracking Global Social Protection Responses to Inflation," Living Paper No. 178842, version 4, World Bank, Washington, DC, 2022.

14 C. Krafft, R. Assaad, and M.A. Marouani, "The Impact of COVID-19 on Middle Eastern and North African Labor Markets," ERF Policy Brief 57, Economic Research Forum, Giza, Egypt, 2021.

15 N.D. Perez, Y. Kassim, C. Ringler, T.S. Thomas, H. ElDidi, and C. Breisinger, *Climate-Resilience Policies and Investments for Egypt's Agriculture Sector: Sustaining Productivity and Food Security*, Food Policy Report (Washington, DC: IFPRI, 2021).

CENTRAL ASIA

1 IMF (International Monetary Fund), *Regional Economic Outlook: Middle East and Central Asia: Mounting Challenges, Decisive Times* (Washington, DC: 2022).

2 National Statistical Committee of the Kyrgyz Republic, *Poverty Rate in the Kyrgyz Republic* (Bishkek: 2022).

3 World Bank, *Europe and Central Asia Economic Update, Spring 2022: War in the Region* (Washington, DC: 2022).

4 IMF, *Regional Economic Outlook: Middle East and Central Asia: Mounting Challenges, Decisive Times* (Washington, DC: 2022).

5 Central Bank of Uzbekistan, *Monetary Policy Survey, Quarter IV, 2022,* in Uzbek (2023).

6 Agency for Strategic Panning and Reforms of the Republic of Kazakhstan, Bureau of National Statistics, accessed Feb. 22, 2023. https://stat.gov.kz/

7 National Statistical Committee of the Kyrgyz Republic, Consumer Price Index in the Kyrgyz Republic, database.

8 FAO/GIEWS (Food and Agriculture Organization, Global Information and Early Warning System), Tajikistan Country Brief, Jan. 3, 2023.

9 S. Sattar, *Addressing the Gender Gap in Europe and Central Asia*, World Bank Knowledge Brief 71408 (Washington, DC: World Bank, 2012); T. Khitarishvili, "Gender Inequalities in Labour Markets in Central Asia," UNDP/ILO conference on Employment, Trade and Human Development in Central Asia; D. Abdurazzakova, K. Kosec, and Z. Parpiev, "Women's Involvement in Intra-household Decision-Making and Infant and Young Child's Feeding Practices in Central Asia," IFPRI Discussion Paper 2058, IFPRI, Washington, DC, 2022.

10 Household, institutional, and societal gender inequalities refer to gender gaps in the following domains: time use and divisions of labor, access and control over resources, intrahousehold decision-making, participation in public life and access to public services, institutional mechanisms for women's empowerment, and human development.

UNICEF, *Gender Counts–Central Asia: A Quantitative Assessment of Gender Inequality and Its Impact on Boys and Girls* (2022).

11 WFP (World Food Programme), *Climate Risks and Food Security in Tajikistan: A Review of Evidence and Priorities for Adaptation* (2017).

12 World Bank, Worldwide Governance Indicators, accessed Jan. 2023.

13 C. Jarvis, G. Pierre, B. Baduel, D. Fayad, A. de Keyserling, B. Sarr, and M. Sumliński, *Economic Governance Reforms to Support Inclusive Growth in the Middle East, North Africa, and Central Asia*, IMF Middle East and Central Asia Department Policy Paper No. 21/01 (Washington, DC: 2021); World Bank, Worldwide Governance Indicators, accessed Jan. 2023.

14 IMF, *Fiscal Monitor: Curbing Corruption* (Washington, DC: April 2019).

15 K. Akramov, K. Kosec, and L. Carrillo, "Covid-19, Rural Poverty, and Women's Role in Decision-Making: Evidence from Khatlon Province in Tajikistan," *OSF Preprints*, available online December 30, 2021.

16 There were widespread reports on social media and daily electronic news outlets about the misallocation of public resources and corruption during the COVID-19 pandemic in the region.

17 WFP, *Food Security Monitoring Update in the Kyrgyz Republic* (Rome: Dec. 2022).

18 OECD (Organisation for Economic Co-operation and Development), *COVID-19 Crisis Response in Central Asia*, Policy Note (Paris: 2020).

19 E. Maltseva, *Kazakhstan's Social Policy Response to Covid-19: Developmentalism and Productive Welfare*, CRC 1342/No.20 Covid-19 Social Policy Response Series (Bremen: Global Dynamics of Social Policy, 2021).

20 World Bank, *Europe and Central Asia Economic Update, Fall 2022: Social Protection for Recovery* (Washington, DC: 2022).

SOUTH ASIA

1 M.O. Nule and Z.L. Xie, "The Road Toward Economic Recovery in South Asia," *World Bank Blog*, Feb. 27, 2023; World Bank, *Poverty and Shared Prosperity 2022: Correcting Course* (Washington, DC: 2022).

2 World Bank, *Global Economic Prospects 2023*, January (Washington, DC: 2023).

3 B. Rother, S. Sosa, D. Kim, et al. "Tackling the Global Food Crisis: Impact, Policy Response, and the Role of the IMF," *IMF Notes* 2022/0004 (Washington, DC: International Monetary Fund, 2022).

4 The World Bank estimates 6.9 percent growth for 2022/23 and 6.6 percent for 2023/24; the Asian Development Bank estimates 7.0 percent and 7.2 percent for those periods..

5 The policy rate reached 2.25 percent between May and December 2022 (Reserve Bank of India, Consumer Price Index–Annual average).

6 J. Alam, "Inflation, Unrest Challenge Bangladesh's 'Miracle Economy'," AP News, Sept. 23, 2022.

7 U. Jamal, "Pakistan's New Budget Aims to Please the IMF," *The Diplomat*, June 15, 2022.

8 World Bank, *Pakistan Development Update: Inflation and the Poor* (Washington, DC: 2022).

9 G.T. Hasnat, M.A. Kabir, and M.A. Hossain, "Major Environmental Issues and Problems of South Asia, Particularly Bangladesh," in *Handbook of Environmental Materials Management* (Springer, 2018).

10 World Bank, *GFDRR Annual Report 2022: Bringing Resilience to Scale* (Washington, DC: 2023).

11 Pakistan, Ministry of Planning Development, and Special Initiatives, *Pakistan Floods 2022: Post-Disaster Needs Assessment* (Government of Pakistan, Asian Development Bank, European Union, UNDP and World Bank, 2022).

12 D.P. Singh, A. Mishra, and P. Shaw, "Taking Cognisance of Households' Inflation Expectations in India," *Reserve Bank of India Working Paper 2*, 2022; A. Dilip and S. Kundu, "Climate Change: Macroeconomic Impact and Policy Options for Mitigating Risks," *RBI Bulletin* 125 (2020): 105–125.

13 World Bank, *Bangladesh Country Climate and Development Report* (Washington, DC: 2022).

14 Indian Tsunami Early Warning System, https://tsunami.incois.gov.in/TEWS/searlywarnings.jsp

15 C. Del Ninno, P.A. Dorosh, and L.C. Smith, "Public Policy, Markets, and Household Coping Strategies in Bangladesh: Avoiding a Food Security Crisis Following the 1998 Floods," *World Development* 31, 7 (2003): 1221–1238.

16 World Bank, "The World Bank in Afghanistan," accessed Feb. 2023.

EAST AND SOUTHEAST ASIA

1 ADB (Asian Development Bank), *Asian Development Outlook 2022 Supplement: Recovery Faces Diverse Challenges* (Manila: 2022).

2 M. Chepeliev, M. Maliszewska, I. Osorio-Rodarte, M.F. Seara, E Pereira, and D. Van Der Mensbrugghe, "Pandemic, Climate Mitigation, and Reshoring: Impacts of a Changing Global Economy on Trade, Incomes, and Poverty," Policy Research Working Paper No. 9955, World Bank, Washington, DC, 2022; UNECE (United Nations Economic Commission for Europe), "Water-Food-Energy-Ecosystem Nexus," accessed December 29, 2022.

3 UN (United Nations), "Transforming Our World: The 2030 Agenda for Sustainable Development," https://sdgs.un.org/2030agenda

4 WFP (World Food Programme), "Hungermap: Asia and the Pacific Insights and Key Trends" (Rome: 2022).

5 FAO (Food and Agriculture Organization of the United Nations), *The State of Agricultural Commodity Markets 2022. The Geography of Food and Agricultural Trade: Policy Approaches for Sustainable Development* (Rome: 2022).

6 FAO, IFAD (International Fund for Agricultural Development), UNICEF, WFP, and WHO (World Health Organization), *The State of Food Security and Nutrition in the World 2022: Repurposing Food and Agricultural Policies to Make Healthy Diets More Affordable* (Rome: FAO, 2022).

7 Development Initiatives, *2021 Global Nutrition Report: The State of Global Nutrition* (Bristol: 2021).

8 A. Elbehri, T. Temel, F. Burcu Ceylan, S. Mittal, D. Kularatne, and D. Dawe, *COVID-19 Pandemic Impacts on Asia and the Pacific: A Regional Review of Socioeconomic, Agrifood and Nutrition Impacts and Policy Responses* (Bangkok: FAO, 2022).

9 ADB, *Southeast Asia Rising from the Pandemic* (Manila: 2022).

10 World Bank, *Global Economic Prospects, June 2022* (Washington, DC: 2022).

11 J-P. Duvieusart, "Emerging Asia Faces a Post-Covid Inequality Trap: Here's How to Avoid It," *World Economic Forum*, May 25, 2022; UNDP (United Nations Development Progamme), *Reducing Inequality in the Decade of Action to Achieve the SDGs and Accelerate Post-Pandemic Recovery* (Bangkok: UNDP Bangkok Regional Hub, 2022).

12 ADB, *Implementing a Green Recovery in Southeast Asia* (Manila: 2022).

13 UN Thailand, "Thailand Economic Focus: Financial Stimulus Packages to Combat COVID-19 and Achieve SDGs," press release, June 8, 2020.

14 A. Elbehri, T. Temel, F. Burcu Ceylan, S. Mittal, D. Kularatne, and D. Dawe, *COVID-19 Pandemic Impacts on Asia and the Pacific: A Regional Review of Socioeconomic, Agrifood and Nutrition Impacts and Policy Responses* (Bangkok: FAO, 2022).

15 M. Basyir, "2023 Budget: Rm1 Billion Parked under Bank Negara Malaysia's Agrofood Financing Scheme," *New Straits Times*, Oct. 7, 2022.

16 D. Eckstein, V. Künzel, and L. Schäfer, *Global Climate Risk Index 2021* (Bonn: Germanwatch, 2021).

17 Fitch Ratings, "China's Drought Unlikely to Unsettle Global Rice Markets," *Fitch Ratings*, Sept. 7, 2022.

18 IPCC (Intergovernmental Panel on Climate Change), *Climate Change 2022: Impacts, Adaptation and Vulnerability. Contribution of Working Group II to the Sixth Assessment Report of the Intergovernmental Panel on Climate Change* (Cambridge: Cambridge University Press, 2022).

19 S. Menegat, A. Ledo, and R. Tirado, "Greenhouse Gas Emissions from Global Production and Use of Nitrogen Synthetic Fertilisers in Agriculture," *Scientific Reports* 12 (2022): 14490.

20 OECD (Organisation for Economic Co-operation and Development), *Agricultural Policy Monitoring and Evaluation 2022* (Paris: 2022).

21 Global Index Insurance Facility, "China Weather Index-Based Insurance Project (China WII)," (Beijing: World Bank Group, 2021).

22 W. J. Barbon, B. Punzalan, R. Wassmann, et al., "Scaling of Climate-Smart Agriculture Via Climate-Smart Villages in Southeast Asia: Insights and Lessons from Vietnam, Laos, Philippines, Cambodia and Myanmar," Working Paper No. 376, CGIAR Research Program on Climate Change, Wageningen, Netherlands, 2021.

23 K. Jones and F. Nti, *Impacts and Repercussions of Price Increases on the Global Fertilizer Market*. (Washington, DC: U.S. Department of Agriculture, 2022).

24 N. Childs and B. Lebeau, *Rice Outlook: October 2022* (Washington, DC: U.S. Department of Agriculture Economic Research Service, 2022).

25 World Bank, *Reforms for Recovery. World Bank East Asia and Pacific Economic Update (October)* (Washington, DC: 2022).

26 Oxford Economics, *The Economic Impact of the Agri-Food Sector in Southeast Asia* (Oxford: 2022).

27 ADB, *Implementing a Green Recovery in Southeast Asia* (Manila: 2022).

LATIN AMERICA AND THE CARIBBEAN

1 E. Díaz-Bonilla, "Democracy and Commodity Cycles in Latin America and the Caribbean," *IFPRI Blog*, Nov. 25, 2019.

2 Economist Intelligence Unit, *Democracy Index 2021: The China Challenge* (London: 2022).

3 E. Díaz-Bonilla and M. Centurion, "Fiscal and Monetary Responses to the COVID-19 Pandemic: Current Conditions and Future Scenarios in Developing Countries," in *COVID-19 and Global Food Security: Two Years Later*, eds. J. McDermott and J. Swinnen, Ch. 27, 162–171 (Washington, DC: IFPRI, 2022).

4 E. Díaz-Bonilla and V. Piñeiro, "Latin America and the Caribbean," in *2021 Global Food Policy Report: Transforming Food Systems after COVID-19*, Regional Developments, 74–105 (Washington, DC: IFPRI, 2021).

5 World Bank, World Development Indicators, 2022.

6 Ibid.

7 World Bank, *The Gendered Impacts of COVID-19 on Labor Markets in Latin America and the Caribbean*, Gender Innovation Lab for Latin America and the Caribbean (LACGIL) Policy Brief (Washington, DC: 2021).

8 Food and Agriculture Organization (FAO), International Fund for Agricultural Development (IFAD), UNICEF, World Food Programme (WFP), and World Health Organization (WHO), *The State of Food Security and Nutrition in the World 2022. Repurposing Food and Agricultural Policies to Make Healthy Diets More Affordable* (Rome: FAO, 2022).

9 For a discussion of policy options for Haiti, see E. Díaz-Bonilla, "Haiti and its Multiple Tragedies: Much More Needs to Be Done," LAC Working Paper 26, IFPRI, Washington, DC, 2022.

10 IPCC (Intergovernmental Panel on Climate Change), *Climate Change 2021: The Physical Science Basis*, eds. V. Masson-Delmotte, P. Zhai, A. Pirani, et al. (Cambridge, UK, and New York: Cambridge University Press, 2021).

11 E. Díaz-Bonilla, "Fiscal and Monetary Responses to the COVID-19 Pandemic: Some Thoughts for Developing Countries and the International Community," in *COVID-19 and Global Food Security*, eds. J. Swinnen and J. McDermott, Ch. 22, 98–101 (Washington, DC: IFPRI, 2020).

12 E. Díaz-Bonilla, *Financing SGD2 and Ending Hunger*, Food Systems Summit Brief (Bonn: UN Food Systems Summit, 2021); E. Díaz-Bonilla, "Using the New IMF Special Drawing Rights for Larger Purposes: Guaranteeing 'Pandemic Recovery Bonds,'" *IFPRI Blog*, October 22, 2021; J. von Braun and E. Díaz-Bonilla, "Letter: Perpetual Bonds Can Help States Fight Hunger," *Financial Times*, June 6, 2021.

13 J. von Braun and E. Díaz-Bonilla, "Letter: Perpetual Bonds Can Help States Fight Hunger," *Financial Times*, June 6, 2021.

14 G20 High Level Independent Panel, *A Global Deal for Our Pandemic Age*, (Financing the Global Commons for Pandemic Preparedness and Response, 2021).

15 C. Andrews, A. de Montesquiou, I. Arévalo Sánchez, et al., *The State of Economic Inclusion Report 2021: The Potential to Scale* (Washington, DC: World Bank, 2021).

Made in the USA
Monee, IL
22 January 2025

10615734R00083